Powerhouse Actin: In Motion

Georgia

Table of Contents

Chapter 1: Introduction

1   The Actin Cytoskeleton

The actin cytoskeleton is an integral part of the cytoskeletal system of all eukaryotes. Actin, a 42kDa protein, is one of the most conserved and abundant proteins found in cells. The actin cytoskeleton is not a static systemic, but highly dynamic: it undergoes active remodeling by processes including polymerization, depolymerization, and association with different actin-binding proteins. The actin cytoskeleton is multifaceted, multifunctional and sustains life in a cell by providing structural support and force generation that drive cell motility, division and shape changes, formation and maintenance of cell junctions, signal transduction (intra and inter-communication), organelle movement, phagocytosis, endocytosis and exocytosis [1–3]. Here, I will describe the structure, function, and dynamics of the actin cytoskeleton in budding yeast and mammals.

1.1   Polymerization of monomeric actin subunits into filamentous actin

Actin protein exists in a dynamic equilibrium between monomeric globular actin (G-actin) and filamentous actin polymer (F-actin) [4]. The crystal structure of monomeric G-actin revealed that it is divided into two lobes with 4 subdomains with a medial ATP-binding pocket known as the nucleotide-binding cleft or ATPase fold that forms between subdomains 2 and 4; and a smaller hydrophobic cleft that mediates interactions between actin and most actin-binding proteins forms between subdomains 1 and 3 (Fig. 1.1a) [5, 6]. In the presence of ATP and cations (Mg^{2+} and Ca^{2+}), multiple monomeric G-actin stack upon each other to form a stable polymer that is polarized with the nucleotide-binding cleft exposed on one end of the polymer [7, 8]. Actin filaments consisting of two actin protofilaments are arranged as a double helix that contains a barbed, fast-growing (+) end and a pointed (-) end (Fig. 1.1b). The inherent polarity of F-actin was first detected by decorating the actin filament with S1 fragment of myosin [9–11].

Polymerization of filamentous actin (F-actin) occurs in 3 phases: 1) nucleation, 2) elongation, and 3) steady-state. The rate-limiting step of actin polymerization is the nucleation, an ATP- and cation-dependent process where more than 1 G-actin monomer come together and form a small stable oligomer called the seed. The seed then acts as a nucleus for subsequent addition of monomeric G-actin to make

F-actin, leading into the elongation phase. Once nucleated, the actin polymer is elongated by further addition of monomeric ATP-bound G-actin on either end of the polymer. Elongation occurs rapidly and when the rates of F-actin formation equal monomeric G-actin levels, steady-state is achieved and there is no change in the total mass of actin filaments during the exchange between G-actin and F-actin.

Interestingly, at steady-state, the rates of actin association and dissociation are kinetically different. This creates different critical concentrations, the concentration at which filaments are formed, on either end promotes the association of G-actin at the barbed end and dissociation of actin at pointed end (Fig 1.1b) [10]. The asymmetrical kinetics of actin polymerization at the barbed and the pointed ends generates a phenomenon known as treadmilling [10]. Treadmilling is powered by cycles of ATP hydrolysis: monomeric G-actin is polymerization competent in the ATP bound form. Once ATP-bound G-actin is incorporated into a filament, the ATP is rapidly (~ 5 sec) hydrolyzed to ADP-Pi. However, Pi is released slowly (~5mins) [8, 12, 13]. Release of Pi from actin-ADP-Pi in actin filaments results in a confirmation change in actin which changes its association/dissociation rates where ADP-actin subunits dissociate faster at the pointed (-) end compared to the barbed (+) end [4]. Thus, at steady state, the actin filament is a mosaic polymer that is enriched in ATP-actin near the barbed end, ADP-Pi actin in the middle, and ADP-actin at the pointed end (Fig. 1.1b) [4]. The ADP-actin that is released from F-actin forms a monomeric pool that is recharged when ADP is exchanged with ATP, creating ATP-actin which can be added to the barbed end of F-actin. Actin-binding regulatory proteins can facilitate this exchange and increase the rates of elongation.

In mammals, treadmilling is evident at the cell's leading edge. Since F-actin in the leading edge is arranged with its barbed end is closer to the cell cortex and its pointed ends are towards the center of the cell, polymerization of F-actin occurs preferentially at the cell cortex, which results in treadmilling of F-actin from the cell cortex toward the interior of the cell. This process, retrograde actin flow, is imperative for cellular motility. In absence of regulatory proteins, treadmilling occurs slowly, thus the rates needed to induced cellular motility require specific regulatory proteins [9].

1.2   Assembly of the actin cytoskeleton

Actin nucleation occurs spontaneously, but slowly because it's intrinsically unfavorable, thus it requires actin-binding regulatory proteins to increase the rate of actin nucleation to levels that support normal physiological behavior [14]. These nucleating promoting factors control the rate and localization of actin nucleation, polymerization and, assembly into networks. Actin can assemble into 2 types of networks: branched filamentous networks mediated by Arp2/3 complex, and unbranched linear F-actin filaments mediated by the formin protein family (Fig. 1.1c).

1.2.1 Arp 2/3 complex-mediated branched filamentous actin network

Arp2/3 complex is a conserved protein complex that nucleates actin into a branched network. It is a 7-unit complex that contains 2 actin-related-proteins, Arp2 and 3, and 5 unique proteins [15]. The Arp2/3 complex has poor nucleation activity, which is activated by nucleating promoting factors and a pre-formed actin filament. Nucleating promoting factors include members of the Wiskott-Aldrich syndrome family protein: WASp, N-WASP (Neural), WAVE (WASP family veprolin homologous protein), and WASH (Wiskott-Aldrich syndrome protein and scar homolog) [16, 17]. Arp2/3 complex nucleating promoting factors are regulated by their protein folding composition, making the c-terminal Arp2/3 activation domain unavailable. WASp contain Rho-GTPase binding domain and an actin monomer binding domain. Binding of Rho-GTPase Cdc42-GTP to WASp binding domain results in a conformation change causing the opening of the previously sequestered Arp2/3 complex activation domain and the binding to actin monomers.

Activated Arp2/3 complex binds to the side of the actin filament and Arp2/3 complex bound actin monomers form a seed to sprout a newly formed barbed end for monomeric G-actin to bind, forming a daughter filament [18]. Binding of Arp2/3 complex to the mother filament and activation of actin polymerization of the daughter filament at that site initiates the formation of a branch at a 70° angle [19]. Organized branching of actin filaments creates a dense, rigid, and stable network.

This actin network is critical for multiple cellular functions. For instance, the dense network produced by the Arp2/3 complex generates a pushing force needed to deform mammalian cell membrane to promote 1) lamellipodium formation during cell migration, 2) phagocytic cup formation during phagocytosis, and 3) comet tails on endosomes that drive endosome movement during endocytosis. In yeast, actin patches, F-actin coated endosomes are mediated by Arp2/3 complex-dependent polymerization branched network formation [4, 20].

1.2.2 Formin mediated polymerization and assembly of unbranched, linear, filamentous actin network

Formin proteins are large (~200kDa), homo-dimeric, multi-domain proteins that are essential and conserved through the entire eukaryotic system [4]. The largest subset of the protein formin family are the *Diaphanous*-related formins, which all share similar domain organization: 1) N-terminal GTPase binding domain (GBD), 2) adjacent DID (Diaphanous inhibitory domain), 3) FH1 (formin-homology 1 domain), 4) FH2, and 5) C-terminal DAD (Diaphanous autoregulatory domain) [21–23]. DID and DAD domains regulate the activity of formins through autoinhibition by interacting with one another, while the FH domains promote actin polymerization by driving elongation of existing actin filaments at their barbed (+) end [24–26]. Crystallography and single-particle EM studies showed that auto-inhibition by the interaction of DID and DAD domains obstructs binding of the C-terminus, FH2 domain to actin, which in turn inhibits formin function in elongation and polymerization of F-actin [24, 25]. Releasing the DID-DAD interactions, concurrently activating formin, is achieved when active Rho-GTPases bind to the GBD domain [22, 23, 27].

The FH2 domain of formins binds with high affinity to the F-actin barbed end [28]. During this process, FH2 domain forms an anti-parallel ring-shaped dimer, which is held together by an α-linker [29, 30]. The α-link is a flexible tether that can expand, thus allowing FH2 ring-like dimeric structure to stretch and encircle actin and/or filament [29]. Both sides of the dimer contain two different actin binding sites, which are marked by conserved residues isoleucine (I1431) and lysine residue (K1601); mutations in these residues result in reduced actin elongation activities [30]. There is evidence that elongation occurs processively, such that the dimeric FH2 domain tracks down the growing barbed (+) end adding subsequent subunits and then dissociates. The mechanism is thought to involve transitioning between an 'open state', actin monomer addition, and a 'closed/capped' state, no monomeric addition [29–36].

Binding of formins to G-actin requires profilin. Profilin is a small abundant actin monomeric-binding protein that contains two separate binding surfaces, one to bind actin monomers and the other to bind to polyproline motifs on the FH1 domain [37]. Profilin is responsible for recycling the nucleotide of the dissociated ADP-actin to ATP. Binding of profilin to ADP-actin results in the opening of the nucleotide-binding cleft, which promotes dissociation of ADP and binding to ATP to yield profilin-ATP-actin complex. Profilin-ATP-actin complex can also bind efficiently to the (+) end and cannot bind to the (-) end, where profilin blocks the addition of G-actin at the − end [4, 38]. After profilin-ATP-actin complex is formed, the profilin complex then binds to the proline regions of the FH1 domain, thus pre-loading actin for addition by FH2. It remains to be fully understood how actin subunits are transferred from the FH1 domain to the FH2 domain. However, it is possible that this occurs by the direct interactions between FH2 and FH1-profilin-actin or at high affinity sites along the FH1 domain [39]. The latter hypothesis suggests profilin-binding sites on FH1 are arranged in a specific order where the highest affinity sites are near the N-terminal, while the lowest affinity sites are nearest to the FH2 domain to help initiate the transfer [40]. Thus, evidence shows the function of profilin is to bind actin subunits and recruit subunits near the FH2 domain, by binding to the FH1 domain, to enhance elongation rates, at least 10-fold, compared to the natural rate of assembly [33, 34]. Profilin is only one among the many actin-binding proteins, which have various regulatory roles in assembly and bundling of actin filaments.

The nature of actin's dynamic assembly contributes to its diverse functions within the cell. The formation of linear, unbranched, F-actin bundles is critical in yeast and mammals. Actin bundles form the structural support for filopodia at the leading edge of motile cells, stress fibers, and focal adhesions. In yeast, actin cables, which function in transporting various cargoes between mother and daughter cells, are formin-dependent actin bundles.

1.3 The actin cytoskeleton of the budding yeast *Saccharomyces cerevisiae*

The yeast actin cytoskeleton is a streamlined version of the mammalian system. Mammals express 6 isoforms of actin ($\alpha_{skeletal}$, $\alpha_{cardiac}$, α_{smooth}, $\gamma_{skeletal}$, β_{cyto}, and, γ_{cyto}) and each isoform is associated with a different structure and function [6]. In contrast, budding yeast expresses only 1 actin gene (*ACT1*), which forms the same structures that are found in the mammalian actin cytoskeleton, including linear and branched F-actin networks [41]. In yeast, actin cables are parallel bundles of F-actin that are generated in part by the formin proteins, Bni1p and Bnr1p (Fig. 1.2a). Branched filaments form actin patches and are nucleated by the Arp2/3 complex and its accessory proteins (Fig. 1.2a). Actin cables and patches are maintained through cellular division and coordinate with each other to initiate and maintain asymmetrical polarized growth of a daughter cell or bud from a mother cell. Thus, these filament-based structures provide the structural basis for polarity, cell morphology, and endocytosis. Similar to mammalians, actin dynamic flow is conserved through the budding yeast. This section will discuss the formation of actin patches and cables in budding yeast.

1.3.1   Actin patch structure and function

Actin patches are endocytic vesicles that are coated with an Arp2/3 complex-dependent branched F-actin network [4, 20, 42]. Patch formation driven Arp2/3 mediated branched network is also regulated by proteins: clathrin and its adaptors, endocytic adaptors, membrane-binding proteins, Arp 2/3 complex and its nucleation promoting factors, protein kinases, and actin-binding proteins (Abp1p, Cap1/2p, Sac6p, Scp1) [4]. These proteins function together to orchestrate distinct steps for patch formation and coordinate the timing and localization of endocytosis with polarized membrane deposition during cellular division.

During the cell cycle, the occurrence of actin patches (i.e. endocytosis) is required for polarized secretion and membrane deposition during yeast cell division. Specifically, in the early G1 phase of the cell cycle, actin patches form at and localize to the nascent emerging bud site in the bud tip [4, 42]. Later in cell division, they form throughout the cortex of the growing bud [4, 42]. Finally, once the bud reaches a size that is similar to that of its mother cell actin patches localize to the bud neck and promote cytokinesis [4, 42]. Actin patches are essential for cell division, most likely because they function in endocytosis and recycling of membrane proteins that are required for polarized site-specific membrane deposition during the yeast cell cycle.

Actin patches are not static structures, rather they are dynamic. In yeast and mammalian cells, the Arp2/3 complex-mediated actin polymerization and network formation drives the movement of endosomes away from the cell cortex. However, in yeast, long distance movement of actin patches use actin cables, another component of the yeast actin cytoskeleton that is described below [4, 20].

1.3.2   Actin cable assembly

Actin cable assembly begins at sites of pooled monomeric G-actin, bud tip and bud neck. At these sites, G-actin is nucleated into F-actin by formin proteins Bni1p at the bud tip and Bnr1p at the bud neck. Formins are further regulated by the Rho GTPases family, which serve as the molecular switches to locally control actin assembly in cells to promote different mechanical functions from whole-cell migration, cell division, and organelle transport. Traditionally, RhoA, Rac1, and Cdc42 are the primary Rho GTPases associated to actin cytoskeleton regulation in mammals [43]. However, yeast express 6 different Rho GTPase: Rho1p, Rho2p, Rho3p, Rho4p, Rho5p, and Cdc42p. All of which have a precise role in regulating specific formins for different functions. As an example, Cdc42p controls formin-mediated, Bni1p, actin assembly at the bud tip during bud emergence, then Rho3/4p regulates actin during bud growth, and Rho1p mediates actin during stress response [44–46].

As described above, profilin is a small conserved monomeric actin-binding protein that interacts with proline-rich regions of the FH1 domain of formins and contains a separate region for binding to monomeric actin subunits. Profilin mediates nucleotide exchange from ADP to ATP on the monomeric G-actin and loading the formin with ATP-bound G-actin, which enhances formin-mediated F-actin polymerization.

Newly polymerized actin filaments are organized into actin cables by the bundling proteins fimbrin (Sac6p) and Abp140p and Bnr1p. Similar to mammals, formin Bnr1p also performs F-actin bundling activity, in addition to its F-actin polymerization activity [4]. Actin cables are further stabilized by tropomyosin proteins, Tpm1p and Tpm2p, which have distinct functions in yeast. Tpm1p functions in stabilization of actin cables by binding to and inhibiting the association of depolymerization/turnover factors like cofilin with actin cables [4, 47]. In contrast, Tpm2p functions in regulating RACF by acting as an inhibitor to Myo1p.

The directionality and length of the actin cables are also orchestrated by actin-binding proteins, including cofilin. Cofilin is a small (~15kDa) conserved protein that binds to the barbed-end grove of ADP-actin subunit and prevents the nucleotide exchange in actin to promote dissociation of actin from filament pointed ends to sever actin filaments [48]. Cofilin mediates F-actin severing by binding to two adjacent ADP-bound actin subunits within F-actin and introducing a twist in the filament at that site to weaken the lateral contacts between the actin subunits [49–52]. Additionally, cofilin and its binding partner, Aip1p, promote rapid turnover of actin cables and patches in yeast by severing tropomyosin-associated actin cables [53]. Thus, the action of cofilin is important for regulating the length of filamentous and turnover, which are needed to have coordination within the actin network to promote movement or other actin functions in the cell.

1.3.3   Actin cable function in cargo transport

Actin cables are bundles of F-actin that align parallel to the mother-daughter axis. The essential function of these structures is to mediate cargo transport and polarized secretion during bud formation and growth (Fig. 1.2b). Actin cables have been implicated as the tracks for anterograde, bud-directed, movement of cargos including mitochondria, secretory vesicles, cortical ER, Golgi, peroxisomes, vacuoles, mRNAs, and spindle alignment elements using forces generated by type V myosin motor proteins Myo2p and Myo4p [4, 54–56]. The finding that type V myosin, which are barbed-end directed motor molecules, drive movement of cargos using actin cables as tracks, indicates that F-actin within actin cables are oriented with their barbed ends in the bud and pointed ends in the mother cell, indicating a retrograde movement.

In addition to anterograde cargo movement, actin cables also mediate retrograde transport of cellular cargos from the bud toward the mother cell, like actin patches [20, 56]. This movement is propelled by retrograde actin cable flow (RACF), a conserved process whereby actin cables undergo movement from the bud toward the mother cell (Fig. 1.2c) [57]. Retrograde actin cable flow is driven, in part, by formins and assembly-mediated translocation of the elongating actin cable away from the bud (Fig. 1.2c) [58]. During this process, newly polymerized F-actin and resident actin cable proteins at assembly sites, bud neck and bud tip, subsequently elongating cables such that the pointed end orients towards the mother cell tip (Fig. 1.2c). In addition to polymerization forces, Myo1p, the type II myosin of yeast that localizes to the bud neck, provides pulling forces for retrograde actin cable flow [58]. This mechanism is conserved in mammalian cells: retrograde actin flow provides the pushing force for retrograde movement of specific cargos (e.g., nuclei, ErbB1 receptors) [59, 60].

1.4   The dynamics of F-actin are critical for many cellular functions

Actin dynamics is a continuous, centripetal flow from actin assembly at the cell cortex and disassembly in the cell body and is conserved from yeast to mammals. This process, retrograde actin flow, affects cell migration and morphology, organelle quality, and lifespan control. Retrograde actin flow is driven by Arp 2/3 complex- or formin-mediated polymerization to form an assembly of actin cytoskeletal meshwork at the cell's edge called lamellipodia or filopodia [3]. Retrograde actin flow protrusive force and direction are controlled by disassembly and actin depolymerization, which is further regulated by Rho GTPases and type II myosin [3, 43, 61, 62].

Two classical structures that are critical for many types of cell migration are lamellipodia and filopodia (Fig. 1.3a). Whole-cell migration requires the concerted effort of both lamellipodium, a flat undulating cellular process, and filopodium a, finger-like cellular protrusions (Fig. 1.2b) [3]. In detail, lamellipodia are planar sheet-like extensions at the cell's leading edge where actin filaments barbed-ends are oriented towards the leading edge and actin is disassembled in the cell body [3, 10, 63]. Lamellipodia generates large protrusion forces that are critical for cell migration, propelling the cell front, which can function as navigation guidance cues and chemotaxis response [3]. Propelling a concerted uniform directionality of a moving cell also requires the formation of the retracting tail, mediated by the linear actin network, at the opposite cellular end [64]. The distinct cellular processes formation, at the front and tail-end of the cell, is regulated by Rho GTPases. For example, RhoA is initiated at the tail end to form the linear branching network at the retracting tail, while Rac1 is at the cell's edge to initiate branching network to form the lamellipodium [43]. In addition, the retractile tail RACF can be further regulated by contractile machinery driven by type II myosin [3, 61, 62]. Moreover, a migrating cell requires rearward movement of intracellular organelles and cargo, which is also driven by retrograde actin flow. For instance, retrograde actin flow drives rearward movement of the nucleus in a migrating cell [60].

Filipodia are dynamic structures that are supported by linear mediated actin bundles. They can function as sensors to probe the extracellular environment during processes including cell migration, neurite outgrowth, and wound healing. Filopodia are also critical for the retrograde movement of proteins, receptors, and viruses that are associated with the plasma membrane of filopodia [59, 65, 66].

Actin dynamics is also required for phagocytosis and intercellular communication. In macrophages, phagocytosis are initiated by binding of proteins on invading microorganisms to cell surface receptors including Fcγ and FcγR [67, 68]. Actin network formation and dynamics are stimulated by receptor activation and generate protrusive forces, which overcome cortical tension and deform the cell membrane to generate phagocytic cups that engulf the invading microorganism [68, 69]. Similarly, actin dynamics that is pivotal for intercellular communication of immune cells during the formation of the immunological synapse. Receptor-mediated interaction of T-cells with antigen-presenting cells triggers extensive cytoskeletal rearrangement, which generates forces to expand the surface area for cell contact and promote receptor clustering at that site [2].

As described above, retrograde actin flow is conserved and occurs during actin cable dynamics in yeast. Our lab discovered retrograde actin cable flow (RACF) and found that manipulating RACF rates has profound effects in mitochondrial quality control and lifespan [57, 70]. Mitochondrial are asymmetrically inherited during asymmetric cell division in budding yeast and mammalian stem cells [71, 72]. In yeast, mitochondria that are more reduced, contain less reactive oxygen species, have higher membrane potential, and therefore higher functioning are preferentially inherited by daughter cells, which promote daughter cell fitness and lifespan. Moreover, we found this is dependent on mitochondrial-cytoskeletal interactions. Since mitochondria use dynamic actin cables as tracks for movement from mother to daughter cells, they are effectively "swimming upstream" against the opposing force of RACF during inheritance. Indeed, we found that increasing rates of RACF promotes the inheritance of higher-functioning, more reduced, mitochondria to daughter cells and enhances longevity [70]. Thus, RACF affects yeast cell fitness and lifespan by acting as a filtering system to prevent low-functioning mitochondria from being inherited by the daughter cell (Fig. 1.3b).

1.5   The unconventional role of the actin cytoskeleton in protein translation

Protein translation is an intricate orchestrated mass assembly of proteins that requires high levels of regulation because it is energetically costly. Recent research identifies the actin cytoskeleton as the system that can control protein synthesis. Global protein synthesis regulation by the actin cytoskeleton is through its effects on the general amino acid control pathway (GAAC) or also known as the integrated stress response in mammals.

1.5.1   The role of F-actin in global control of protein synthesis

Translation is divided into 4 major steps: initiation, elongation, termination, and ribosome recycling. Translational initiation is regulated by elongation initiation factor family members. During this process, a GTP-bound form of eIF2 translation factor binds tRNA^{Met} to form a ternary complex (TC) and delivers it to the small 40S ribosomal subunit to form a 43S complex. The 43S complex scans the mRNA to find its corresponding start codon, which triggers binding of 60S ribosomal subunit to form 80S ribosome and initiate translation. eIF2 is a heterotrimeric complex consisting of 3 subunits: α, β, and γ. Once eIF2 completes the transfer of tRNA^{Met} to the A-site of the ribosome upon pairing with start, Met, codon, GTP becomes hydrolyzed. Recycling GDP to GTP to reinitiate eIF2 function is controlled by eIF2B, a eIF2 guanine nucleotide exchange factor (GEF) that binds to eIF2β [73]. Whereas eIF2α is a regulatory domain that is phosphorylated by the kinase, Gcn2, in response to amino acid starvation and halts global protein synthesis by inhibiting eIF2B GEF activity [74].

Translation elongation occurs once initiation is completed. During translational elongation, delivering amino-acetylated tRNA (aa-tRNA) is controlled by an aminoacyl-tRNA transferase, eEF1A. In its GTP-bound form, eEF1A transfers the correct aa-tRNA to the ribosome A-site to continue the elongation of the polypeptide. eEF1A is an abundant translation elongation factor that is conserved among eukaryotes. Interestingly, eEF1A has roles beyond its canonical role in translation elongation, including functions in control of the cell cycle, actin bundling, and apoptosis [75].

Initial isolation of eEF1A from slime mold, *Dictyostelium discodieum,* revealed that it is an actin-binding protein [76]. Most interestingly, eEF1A-actin cytoskeleton interactions are evolutionary conserved where eEF1A from multiple organisms can bind to and bundle actin filaments [76, 77]. Cross-linking actin filaments by eEF1A distinctly excludes other actin-binding proteins from cross linking actin filamentous [78, 79]. Overexpression of eEF1A results in disorganized actin cytoskeleton in budding yeast [79–81]. Detailed analysis of the eEF1A protein composition in its actin function have found that 1) mutation of domains 2 and 3 result in actin disorganization, 2) mutations in domain 2 have no effect on protein synthesis rates, and 3) domain 3 mutations result in translational initiation defects and increase phosphorylation of eIF2α [79, 81]. However, binding of eEF1A to actin decreases its affinity for guanine nucleotide, increasing the rate of GTP hydrolysis, and reduced translation [82]. This is not a unidirectional effect, but bi-directional: mutations in actin-binding proteins, like yeast Tpm1p and Bnip/Bnr1p, exhibit reduced translation and/or 80S ribosome accumulation [79]. Thus, there is a link between two distinct cellular process, cytoskeleton and translation, and eEF1A as the central protein that connects the two processes.

1.5.2   Role for G-actin in control of translation

Recent research supports a role for G-actin for regulating protein synthesis through Gcn2 activation, a serine/threonine protein kinase involved in modulating protein synthesis in response to nutrient deprivation. Research found eEF1A has a Gcn2 binding domain at its C-terminus for Gcn2 and only when experiencing nutrient starvation initiates the dissociation between eEF1A and Gcn2 [83]. Thus, eEF1A binds to and sequesters Gcn2, preventing Gcn2 activation under basal conditions [83].

Gcn2 activity is further regulated by Yih1p, a yeast ortholog of IMPACT. Research has uncovered that Yih1p inhibits Gcn2 activation by competing with Gcn2p for binding to Gcn2 activator, Gcn1p [84]. Moreover, Yih1p can form a complex with G-actin and binding inhibits Yih1p interaction with Gcn1p, thus, interaction between actin and Yih1p can regulate Gcn2p activation [84, 85]. Mammalian homolog, IMPACT, of yeast Yih1p exhibits the same function; binds to G-actin and overexpression of IMACT impairs Gcn2 activation [84, 86–88]. Interestingly, the induction of actin depolymerization triggers Gcn2 activation and concomitantly results in reduced translation, suggesting G-actin sequesters Yihp/IMPACT [88].

Overall, this evidence implies that under conditions favoring F-actin assembly and low G-actin levels, Gcn2 remains in an inactive state because Gcn1 (Gcn2 activator) is sequestered by Yihp/IMPACT and Gcn2 is sequestered by eEF1A. In contrast, actin depolymerization displaces eEF1A from Gcn2 causing its activation and triggering translation inhibition [88]. Depolymerization of actin increases G-actin levels, which results in more sequestration of Yihp/IMPACT by G-actin, thus, promoting the interaction Gcn1-Gcn2, activating Gcn2 to reduce global protein synthesis. However, as described earlier, eEF1A functions to bundle actin filaments. It remains unclear how F-actin binding and bundling activity by eEF1a affects this mechanism for G-actin control of Gcn2 activation and its global protein synthesis control. Nevertheless, evidence show actin cytoskeleton is involved in regulating protein translation and proposes actin as a nexus between translation and nutrient sensing.

## 2	Branched-chained amino acids

Branched-chain amino acids (BCAAs) are essential amino acids in higher eukaryotes. Moreover, defects in BCAA homeostasis result in an array of metabolic diseases. Indeed, high levels of BCAAs are toxic and are associated with the development of type 2 diabetes, insulin resistance, liver disease, and neurological defects [89–91]. Finally, although defects in BCAA can reduce lifespan, it is not clear whether depletion or supplementation of BCAA can extend lifespan. Therefore, understanding the BCAA metabolism is critical to understanding its clinical manifestations and impact on the aging process.

### 2.1	BCAA synthesis and metabolism in yeast

Although BCAAs are essential in higher eukaryotes, budding yeast can synthesize all amino acids, including essential BCAAs, leucine, isoleucine, and valine. BCAA biosynthesis commences from a common pathway that metabolizes 2 pyruvates into valine or 2-ketobutyrate (a threonine derivative) to isoleucine and intermediates in valine biosynthesis are used as precursors for leucine production. These synthetic processes occur in mitochondria and the cytoplasm.

The synthesis of BCAAs is not complex (Fig 1.4). The first step common in all BCAAs is conversion of either 2 pyruvate or 2-ketobutyrate to form 2-acetolactate (a valine and leucine precursor) or 2-acetohydroxyburyrate (an isoleucine precursor) by acetolactate synthetase, Ilv2p and Ilv6p [92, 93]. Subsequently, these acetohydroxyacids are reduced and isomerized to 2,3-dihydroxy branched acids in reactions catalyzed by ketol-acid reductoisomerase, Ilv5p. They are further dehydrated by dihyroxyacid dehydratase, Ilv3p, to form α-keto intermediates (2-ketosiovalerate or 2-keto-3-methyvalaerate), which are then converted to L-valine or L-isoleucine, respectively, by either of branched-chain transaminases (BAT) paralog proteins, Bat1p and Bat2p. 2-ketoisovalerate can be channeled from the valine to leucine biosynthesis in reactions catalyzed by the α-isopropylmalate synthase paralog proteins, Leu4p and Leu9p, to form $2(\alpha)$-isopropylmalate. Interestingly, studies show Leu4p and Leu9p, are regulated by different proteins during leucine biosynthesis [94]. α-isopropylmalate is then exported out of the mitochondrion and converted to leucine by 3 additional steps in reactions catalyzed by 3-isopropylmalate isomerase (Leu1p), 3-isopropylmalate dehydrogenase (Leu2p), and by either Bat1p or Bat2p. Lastly, α-isopropylmalate is a stress response signal that accumulates during leucine starvation and is transported to the nucleus where it activates the zinc-knuckle transcription factor, Leu3p [92]. Leu3p is both a transcriptional repressor and activator responsible for regulating BCAA biosynthetic genes, like the BAT proteins [92].

BAT proteins are important enzymes for the last and first steps of BCAA biosynthesis and catabolism. Bat1p and Bat2p are paralog proteins that show 81% identity and were retained during the whole-genome duplication event in *S. cerevisiae*. While these are paralog proteins, they evolved to have different functions, localization, and expression in BCAA homeostasis [95–97]. Such that Bat1p preferentially mediates BCAA synthesis, while Bat2p preferentially mediates BCAA catabolism. Furthermore, their function and expression occur in a compensatory manner, meaning when one is activated/expressed the other is repressed, according to its nitrogen status [97]. For instance, activation of Leu3p by glutamine depletion leads to *BAT1* expression, while repressing *Bat2*, to promote BCAA synthesis [97]. On the other hand, *BAT1/2* expression can be induced by the transcriptional activator, Gcn4p, in response to amino acid deprivation [97]. BAT1 and BAT2 expression are high and low, respectively, during mid-log phase. In contrast, *BAT2* expression is high and *BAT1* is repressed in a stationary phase yeast [96]. Moreover, Bat1p has a mitochondrial-targeting signal and resides at the mitochondrion and Bat2p localizes in the cytoplasm [96]. Recent studies suggest the mitochondrion is the major site for valine biosynthesis because valine synthesis preferentially uses Bat1p, rather than Bat2p, which localizes to mitochondria [98]. Therefore, the last step of BCAA synthesis can either take place at the mitochondrion or in the cytosol through the action of either Bat1p or Bat2p [96].

2.2 Sensing and controlling BCAA levels

Cell viability during nutrient limitations is determined by the cellular response to that environmental cue. In simple organisms, like yeast, sense carbon and nitrogen availability to adjust their metabolism and anabolism processes accordingly. In particular, amino acids are nitrogen sources that are sensed by multiple pathways, including the SPS, GCN2, and target of rapamycin complex 1 (TORC1) pathways [99]. This section focuses on 2 fundamental molecular pathways that coordinate amino acid availability with anabolic processes, starvation response and catabolic processes, TORC1, and general amino acid control (GAAC) pathways.

2.2.1 BCAAs controls TORC1 activity

TORC1 is a conserved master regulator that senses intracellular amino acid availability and regulates fundamental processes involved with cellular growth, such as translation, ribosomal biogenesis, protein synthesis, and autophagy in response to that nutritional cue [100, 101]. TOR, a phosphatidylinositol kinase-related kinase, was initially discovered in budding yeast in a mutagenesis-based screen for resistance to the growth inhibitory effects by rapamycin [102, 103]. As a result of whole genome duplication, yeast have two *TOR* genes, *TOR1* and *TOR2*. This distinguishes yeast from mammals, which have only one TOR gene and protein, mTOR. Nonetheless, all Tor proteins share the same domain features: N-terminal HEAT (Huntington, elongation factor 3, regulatory subunit A of PP2A, Tor1) repeats, and FAT or FRB (FKBP12-rapamycin-binding) domains within the protein and at its C-terminus [104]. Tor proteins are incorporated into two distinct protein kinase complexes, TORC1 and TORC2. TORC1 is rapamycin-sensitive and regulates protein synthesis, ribosome biogenesis, transcription, cell cycle progression, nutrient uptake, and autophagy. When rapamycin interacts with Fpr1p, yeast homolog FK506-binding protein 12, the complex then binds to the FRB domain of TOR and inhibits its activity [105]. In contrast, TORC2 is rapamycin-insensitive and regulates the actin cytoskeleton organization, endocytosis, lipid synthesis, and cell survival [106]. Below, I describe the TORC1 pathway of yeast.

The TORC1 complex consists of Tor1p or Tor2p and HEAT repeat domain-binding proteins: Kog1p, Lst8p, and Tco89p [107–109]. Kog1p, the yeast ortholog of the mammalian mTOR protein Raptor, recruits Tor1 substrates and regulates Tor1p activity [107–109]. Studies show the conserved Lst8p binds to TOR kinase domain and positively regulates TOR catalytic activity, however, its precise role is unknown [100, 107, 110, 111]. Tco89p is a non-essential TORC1 component that regulates TORC1 activity and is suggested to help tether TORC1 to the vacuolar membrane [109, 112, 113].

TORC1 permanently resides at the vacuolar membrane (yeast lysosome), unlike mammals, and is regulated by small Rag GTPases, Gtr1p and Gtr2p, which are yeast homologs of mammalian RagA/B and RagC/D, respectively (Fig. 1.5) [112, 114]. Gtr1p and Grt2p form a heterodimer that is tethered to the vacuolar membrane by the EGO complex, a palmitoylate and myristoylated complex made up of Ego 1-3 [112, 115–117]. When amino acids are abundant, Gtr1p and Gtr2p, are active and in their GTP and GDP bound forms, respectively. Active Gtr1/2p tightly binds to Kog1 and Tco89 to stimulate TORC1 activity [112, 118].

However, when amino acids, like leucine, are not available, the SEACIT complex (Seh1-associated sub-complex inhibiting TORC1) functions as a GTPase activating protein (GAP) for Gtr1p, which induces Gtr1p GTPase activity and converts Grt1p-GTP to its inactive GDP-bound form [119, 120]. Binding to GDP induces a conformational change in Gtr1p, which releases Gtr1/2p from TORC1 and inhibits TORC1 signaling [112, 121]. SEACIT complex is negatively regulated by binding to SEACAT (Seh1-associated complex subcomplex activating TORC1) [120, 122]. Moreover, the vacuolar protein Vam6p is a guanine nucleotide exchange factor (GEF) for Gtr1p [112]. Lst4p and Lst7p are GAPs for Gtr2p, although, the identity of Grt2p GEF remains unsolved [123].

Activation of TORC1 kinase activity results in the induction of downstream pathways that promote cell growth and protein synthesis, while inactivation of TORC1 triggers catabolic processes and/or stress responses. Some main downstream effectors of TORC1 are the protein kinases Sch9p and Ypk3p, Tap42-PPase complex, and transcription factor Sfp1 [101, 124]. Phosphorylation of Sch9p by active TORC1 activates Sch9p kinase activity and phosphorylate Rim15p, which promotes entry into G_0 phase of the cell division cycle [125]. Independent to its cell cycle progression role, phosphorylated Sch9p also involved in translation by inducing the expression of ribi genes expression, genes involved in ribosomal biogenesis [101, 126]. Ribi and ribosomal protein gene expression is further regulated by Sfp1 during active TORC1 activity [101, 127]. Phosphorylation Ypk3p by active TORC1 results in active Ypk3p phosphorylation of ribosomal protein S6 to promotes protein synthesis [128]. TORC1 effector Tap42-PPase complex relays TORC1 activity to control cell wall integrity, cell cycle progression, the expression of nitrogen catabolite repressed genes, and stress response genes [101, 129–132].

Branched-chain amino acids (BCAAs) are known TORC1 regulators. Specifically, elevated levels of the BCAA, leucine, activate TORC1 through effects on the Rag GTPase, Gtr1p. Leucyl-tRNA synthetase (LeuRS), a conserved cytoplasmic leucine sensor, activates TORC1. When leucine binds to LeuRS initiates favors binding to Gtr1p in its GTP bound form [133]. It is not clear whether SEACAT has a role in this pathway for leucine regulation of TORC1 [134]. In yeast,. BCAA transaminase proteins Bat1p and Bat2p effects on intracellular leucine levels also contribute to controlling TORC1 activity through LeuRS and the

EGO complex [135]. Interestingly, additional leucine sensors that regulate TORC1 activity such as SESTRIN proteins and leucine transporters (SLC7A5-SLC3A2) have been identified in mammalian cells [136, 137]. It will be interesting to see if these sensors are also present in yeast.

2.2.2 BCAAs depletion activates the general amino acid control (GAAC) pathway

The GAAC and TORC1 have complementary functions in the cellular response to nutrient availability. TORC1 up-regulates translation, cell division, and down-regulates autophagy and stress response when nutrients are available. In contrast, the general amino acid control (GAAC) pathway drives changes in gene expression to adapt to amino acid depletion conditions. Specifically, the GAAC signaling pathway GAAC down-regulates translation and up-regulates autophagy and the biosynthesis of amino acid, purine, and other metabolites [138].

2.2.2.1 Translation regulation of GAAC

The GAAC kinase Gcn2p down-regulates protein synthesis by inhibiting ternary complex (TC) formation during translational initiation when amino acids are limiting. The target for this regulatory event is the eIF2B also known as the general control repressible protein complex Gcd1/2/6/7/11p, which is a GEF for eIF2 [138]. Translation initiation begins when eIF2 GTP-bound form binds to tRNA^{Met} to form the TC. During the base-pairing of the 43S complex, complex formed of the interaction between TC and 40S ribosome, with the start anticodon: eIF2-GTP is hydrolyzed to its inactive GDP-bound form. eIF2-GDP is then recycled back to GTP-bound form by eIF2B. Accumulation of deacetylated, uncharged tRNAs in response to amino acid limitation results in activation Gcn2p, which leads to Gcn2p-catalyzed phosphorylation of the α subunit of eIF2 [138, 139]. This, in turn, prevents eIF2 interaction with eIF2B by inhibiting its GEF activity, thus eIF2 remains bound to its inactive GDP form and prevents TC complex formation [138].

The mechanism whereby Gnc2p senses accumulation of uncharged tRNAs is not known, but there are numerous studies suggesting hypothetical mechanistic routes. Gcn2p protein structure is organized into roughly 5 main domains: an N-terminus is the Gcn1p binding domain, a pseudokinase domain of unknown function, a kinase domain, an HisRS-like domain that binds uncharged tRNAs and a C-terminal domain [138]. Gcn2p protein is autoinhibited through intramolecular interactions mediated by its pseudokinase domain. Binding of the uncharged tRNA at the HisRS domain relieves the autoinhibition, which activates the Gcn2p kinase domain and phosphorylates eIF2α [140–145]. Binding of Gcn1p, a Gcn2 effector protein, to Gcn2p at its N-terminal binding domain increases the phosphorylation rate of eIF2α by Gcn2p in amino acid-starved cells [138]. There is evidence that Gcn2p sense uncharged tRNAs that are bound at the aminoacyl acceptor site (A-site) of the ribosome and that Gcn1p is directly involved in transferring the uncharged tRNA starvation signal to Gcn2p [146]. However, it is not clear if Gcn1p serves as a scaffold protein to bring Gcn2p closer to the ribosome A-site and/or as a mediator that promotes binding and delivering of the uncharged tRNAs to that site [146–148].

2.2.2.2 Transcriptional regulation of GAAC

Activated GAAC results in translation of a transcriptional activator, Gcn4p (General Control Nonderepressible). It is a transcriptional activator that controls the transcription of over 500 genes, which include more than 30 amino acid biosynthetic genes, various aminoacyl-tRNA synthetases, and pathway-specific activators [138]. Amino acid limitation results in increased Gcn4p levels by GAAC-mediated derepression of Gcn4p translation in yeast.

GCN4 mRNA contains mRNA leader of 4 additional upstream short ORFs (μORF) that are 2 to 3 codons in length and act as translation barriers. Although deletion of all μORFs in *GCN4* leads to constitutive activation of Gcn4p translation, μORFs 1 and 4 are the primary mediators of *GCN4* translational control [138]. μORF1 is the weak translational barrier resulting in a "leaky" scanning and μORF4 is the critical negative regulator of Gcn4 translation [138]. During normal translation, each of the 4 μORFs upstream of the *GCN4* coding region are translated and ribosomes then dissociate. In normal conditions, the TC concentrations are high, eIF2 nucleotide is recycled by eIF2B, thus 40S subunit quickly rebinds a new TC in time to reinitiate at μORF4, the strongest inhibitory translation barrier, and completely dissociates the ribosome complex upon translation before Gcn4 mRNA can be translated (Fig. 1.6a). Under amino acid starvation conditions where the TC formation (Met-tRNA$_i^{Met}$, GTP and eIF2) is low due to the Gcn2p-inhibition of recycling of eIF2 to its GTP bound state, 40S ribosome subunits that do not bind to TC scan past μORFs through its leaky barrier and initiate translation at the GCN4 start codon (Fig. 1.6b) [138]. As a result, the translation of Gcn4p is derepressed, inducing the transcription of amino acid biosynthetic genes, and autophagy-related genes, which in turn, promotes survival under amino acid limiting conditions.

3   The biological mechanisms of aging

The population prediction from the World Health Organization shows that by 2050 the majority of the human population will be aged 65 and over. This will have a considerable impact on our medical systems by increasing healthcare costs and demand for more physicians due to the rise in age-related health diseases and challenges of treating individuals with co-morbidities. Therefore, understanding the biology of aging and how it contributes to age-related pathologies are crucial for the future.

3.1   Yeast as an aging model organism

The mechanisms underlying lifespan control are studied in model single-cell organisms, like yeast, to multicellular organisms. While each model organism has their own benefits, the aging process is more complex in "higher" the eukaryotic species. The budding yeast, *Saccharomyces cerevisiae*, divides rapidly, has a short lifespan (60 hrs), is genetically manipulatable, and is amenable to high-throughput analysis. Moreover, yeast can be used to study 2 forms of aging: chronological and replicative lifespan.

Chronological lifespan (CLS) is a model for aging of terminally differentiated cells (e.g. neurons or muscle fibers). This model studies yeast that are exposed to nutrient-limited conditions. As a result, yeast cells remain viable in a non-dividing, quiescent state for a period of time known as the chronological life span (CLS) (Fig. 1.7) [149]. In contrast, replicative lifespan (RLS) is a model for aging in cells that are cell division-competent (e.g. stem cells or epithelial cells). During yeast cell division, a bud emerges at a specific site on mother cells, grows, and separates to yield a smaller daughter cell and a larger mother cell. Mother cells divide a finite number of times and the number of daughter cells produced by a mother cell is known as the replicative life span (RLS) (Fig. 1.7) [150].

In yeast, as in humans, babies are born young largely independent of the age of their parents. The finding that this process, mother-daughter age asymmetry, occurs in single-cell organisms like budding yeast, led to the model that aging determinants are asymmetrically inherited during yeast cell division. Indeed, yeast mother cells retain aging factors (including oxidatively damaged proteins and lower function organelles), while daughter cells preferentially inherit rejuvenating factors including higher functioning mitochondria and vacuoles, antioxidant activity [70, 71, 151, 152]. Moreover, recent studies indicate that mitochondria are asymmetrically inherited during asymmetric cell division of human mammary stem-like cells and that this process affects cell fate [72]. Thus, RLS is also a model to study mechanisms underlying mother-daughter age asymmetry.

In actively dividing mid-log phase yeast, the vast majority of cells are young because old cells are constantly diluted with every emerging bud, or daughter cell, from mother cells. Therefore, methods to monitor aging in single cells and to isolate yeast of different replicative age are essential to use yeast as a model for aging. The original method that remains the gold standard for determining RLS in yeast relies on micromanipulation and microdissection to separate newly developed daughters from the mother cells and count the number of daughter cells that are produced from a single mother cell [153, 154]. Recently, microfluidic chambers have been developed that trap individual mother cells, remove newly developed daughter cells, monitor labeled protein or biosensors during aging, and are designed for high throughput live-cell imaging analysis [155, 156].

Methods to isolate cells at a defined age class have also been instrumental for studying the aging process in yeast and for employing systems-based approaches to study genetic, epigenetic, transcriptional, and proteomic changes in yeast as they age [156]. One approach to isolate cells as a function of age that was developed decades ago and remains the foundation for current methods relies on labeling the cell wall of mid-log phase cells, which are predominantly young, with biotin, propagating those cells to allow labeled cells to age, and use affinity purification to separate biotinylated old cells at a defined replicated age range from unlabeled young cells (Fig. 1.7) [157]. A modification of this method was developed recently in which biotinylated yeasts are maintained in mid-log phase in mini-bioreactors using streptavidin-coated magnetic beads, young daughter cells are removed by a constant flow of fresh media into the system, and cells of defined replicative age are harvested by removing cells from magnet in the bioreactor (Fig. 1.7) [158]. With these technologies combined with the power of yeast genetics and genomics has allowed our lab and others to study biological mechanisms underlying the lifespan control.

3.2   Biological hallmarks of aging

There are 9 hallmarks of aging [159]. Here, I will focus on the aging hallmarks that are conserved in single-cell organisms: genomic instability, epigenetic alterations, loss of proteostasis, and deregulated nutrient sensing [160]. Genomic instability, or the unintended alterations of the genome, is a common denominator of aging that is due, in part, to the accumulation of damage throughout life [159]. Premature aging in Werner and Bloom syndrome are consequences of increased DNA damage [161]. Indeed, deletion of *SGS1*, the yeast homolog of the RecQ helicase that is the target for mutations in Werner and Bloom syndrome, results in a profound decrease in lifespan [162]. Aging organisms also experience epigenetic alterations including changes in DNA methylation, post-translational modification of histones, and chromatin remodeling changes [159]. The relevance of histone modification for the aging process is evident from analysis of the Sirtuin family of proteins, lifespan modulators, and NAD-dependent protein deacetylases that catalyze histone deacetylation. Deletion of *SIR2* in budding yeast reduces lifespan, while overexpression extends lifespan [163]. Similarly, mice deficient in SIRT6 undergo accelerated aging and overexpression of SIRT6 results in extended lifespan [164, 165]. The remaining conserved hallmarks are described in detail in the following sections.

3.2.1   Age-associated decline in mitochondria function

Mitochondrial dysfunction during aging has been studied for over a century. Early studies revealed that mitochondria are central mediators of aging through their function in reactive oxygen species (ROS) production [166, 167]. However, it is now clear that ROS has multiple effects on cell fitness and lifespan. Although mitochondrial ROS production increases with age and high levels of ROS are damaging, exposure to low levels of ROS can induce stress resistance and promote lifespan (i.e. hormesis) [159]. Moreover, premature aging can be produced by defects in mitochondria without any increase in ROS production by the organelle: mice bearing mutations in mitochondria DNA polymerase γ exhibit accumulation of mutations in mtDNA, defects in mitochondrial function, and reduced lifespan with no obvious increase in mitochondrial ROS production [168, 169]. Thus, while there are clear links between mitochondria and lifespan control, there may be multiple mechanisms whereby mitochondria contribute to and affect the aging process.

Mitochondria in budding yeast also undergo age-associated declines in membrane potential and redox state [71]. Moreover, promoting inheritance of higher functioning mitochondria by yeast daughter cells extends lifespan and promotes healthspan [70]. Other studies revealed a progressive loss in acidification of the vacuole (lysosome) with age, which impacts vacuolar function in amino acid storage. These studies also revealed that reduced mitochondrial function with age is driven in part by the deterioration of vacuolar acidification: overexpression of the vacuolar ATPase, Vma1, restores vacuolar pH, mitochondrial function and extends lifespan [151]. Thus, there is an interplay between mitochondria, nutrient sensing pathways, and lifespan control.

The retrograde response pathway, a signaling pathway for communication between mitochondria and the nucleus that is triggered by defects in mitochondrial function, may link mitochondria, nutrient sensing, and aging. Activation of the retrograde response pathway results in changes in nuclear gene expression that promotes mitochondrial energy production and homeostasis as well as cell survival [170]. Moreover, induction of retrograde signaling was proven to increase lifespan in yeast, worms, flies, and mice [171–173]. Rtg1p and Rtg3p are transcription factors that relocalize from the cytosol to the nucleus, where they activate the expression of nuclear-encoded mitochondrial genes, in response to activation of the retrograde response pathway. Interestingly, the localization of Rtg1/3p is regulated by the nutrient sensing, lifespan regulating TORC1 pathway. Specifically, treatment with rapamycin inhibits TORC1 and activates the retrograde response pathway [174, 175]. My studies revealed a novel mechanism for nutrient control of the actin cytoskeleton that impacts mitochondria and lifespan that is not dependent upon TORC1.

3.2.2   Role for nutrient sensing pathways in lifespan control

Dietary restriction, reduction of food intake without malnourishment, is an intervention that can prolong lifespan and delay age-associated pathologies in all organisms tested (yeast, nematodes, fruit flies, mice, and primates). Interestingly, longevity produced by dietary restriction is effective when administered after development [176–178]. Thus, extension lifespan and healthspan by dietary restriction can be achieved without the cost of delayed development. The nutrient sensing pathways TORC1 and GAAC are major contributors to lifespan extension produced by dietary restriction. Below, I describe the role of these nutrient sensing pathways in lifespan control.

3.2.2.1   Lifespan control through TORC1

Dietary restriction of glucose or amino acids (as opposed to reducing calories, per se) increases lifespan, along with rapamycin treatment [179, 180]. These two interventions specifically target the TORC1 activity. Deletion of *TOR1* in yeast is sufficient to extend RLS and no additive effects were applied under caloric restriction (CR), implying CR works by targeting the TORC1 [181]. Finally, lifespan extension by genetic manipulation of the TOR pathway, dietary restriction, or rapamycin treatment is conserved amongst eukaryotes [181–186]. TORC1 affects lifespan through effects on translation and autophagy. Here, I describe how TORC1 effects on protein synthesis contribute to lifespan control.

Protein synthesis is energy intensive. TORC1 signaling is one mechanism to down-regulate translation in response to nutrient limitations, which in turn delays aging. In mammals, mTORC1 affects translation by activating ribosomal S6 protein kinase (S6K) and by repressing the translational inhibitor 4E-BP. 4E-BP binds to eIF4E directly and precludes interaction of eIF4E with the cap structure on mRNA during assembly of ribosome initiation complex on mRNA. Inhibition of TORC1, by genetic manipulation, rapamycin treatment, or dietary restriction, allows 4E-BP to suppress translation from yeast to mammals [187–190].

Translation control driven by TORC1 is conserved and can be observed in yeast. In yeast, Sch9p is a direct target of TORC1, which is responsible for regulating translation by controlling the expression of ribosomal proteins. Thus, deletion of *SCH9* extends lifespan [181]. Therefore, TORC1 in all eukaryotes affects lifespan, in part, by dampening translation in response to nutrient limitation.

Thus, regulating protein translation is key for lifespan control. Components of the translation initiation complex, including eIF4A (yeast TIF1/2) and eIF4G (yeast TIF4631), also extends lifespan in yeast [191]. Deletion of ribosomal proteins *RPL31a* and *RPL6*, other ribosome genes (*RPL10*, *RPS6B*, or *RPS18A/B*) or gene encoding components of the 60S ribosomal subunit results in lifespan extension in yeast [181, 192, 193]. Indeed, defects in ribosomal assembly and reduced translation in the ribosomal mutants contributes to this longevity phenotype [194].

3.2.2.2   Lifespan control through GAAC

Another nutrient sensing pathway that regulates ribosomal assembly and global translation is the general amino acid control (GAAC) pathway. Activating the GAAC pathway through Gcn2p phosphorylates eIF2α causing a global reduction in protein synthesis. Increasing the activation of GAAC and reducing translation increases the replicative lifespan in yeast [180]. In addition, analysis of long-lived ribosomal 60S mutants, previously described above, revealed that lifespan extension is mediated by GAAC component,

Gcn4p. Specifically, deleting *GCN4* blocks the lifespan extension produced by mutation of ribosomal 60S subunit components [176]. Moreover, *GCN4* overexpression is sufficient to increase the replicative lifespan of a yeast cell [177, 180]. Interestingly, *GCN4* overexpression activates the autophagy flux and extends RLS independent of TORC1 [180]. Thus, reducing translation is a well-established mechanism by GAAC pathway, which can extend lifespan in response to nutrient limitation in yeast (Fig. 1.8).

3.2.3   Losing protein quality control with age

Proteostasis is a protein quality control system that maintains the equilibrium between synthesis, conformational maintenance, and degradation of damaged proteins through pathways that regulate protein production, autophagy, proteasomal degradation, and chaperone-mediated protein folding [159, 195]. Deficient proteostasis contributes to aging-related pathology, including Alzheimer's disease, Parkinson's disease, and cataracts [196]. Heat shock proteins are the most prominent chaperone family that functions in stabilizing the correct protein fold and are stress factors that are rapidly induced in response to elevated temperatures and other stress stimuli [197]. During aging, stress-induced synthesis of heat shock proteins is impaired. Moreover, increasing the induction of chaperone proteins results in an increase in lifespan in both uni-and multicellular organisms [197]. For instance, activation of the transcription factor HSF-1, a master regulator of the heat-shock response, increases thermotolerance and lifespan in nematodes [198, 199]. Moreover, recent evidence supports a link between cellular proteostasis and metabolic processes, a key homeostatic mechanism [195].

Promoting proteostasis by chaperone expression in budding yeast extends lifespan, reduces TOR signaling, and promotes mitochondrial biogenesis [195]. This interconnected relationship can also be applied to protein degradative pathways of proteostasis, like autophagy. Autophagy is a cellular recycling process that degrades cytoplasmic organelles and macromolecules to promote cell survival and maintenance [200]. There are 3 forms of autophagy marcoautophagy, microautophagy, and chaperone autophagy. It has been recorded different mammalian tissues show a decline in macroautophagic activity with age [201, 202]. Moreover, increasing autophagic levels by overexpressing the autophagy-related protein, ATG5, in mice extends lifespan [203]. Interestingly, knocking down autophagy-related genes of worms blocks the lifespan extension produced by mutating *daf-2*, insulin-life growth factor [204, 205].

25

TORC1 activity regulates macroautophagy induction is a conserved mechanism: active TORC1 in nutrient-rich conditions inhibit autophagy, while TORC1 activity inhibition by nutrient-limiting conditions or rapamycin activates macroautophagy [206–208]. Reduced TORC1 activity by rapamycin extends lifespan and promotes healthspan in mice [182]. Finally, autophagic genes are regulated by Gcn4p, transcription factor induced by GAAC pathway. The overexpression of *GCN4* extended replicative lifespan in yeast is dependent on inducing the autophagy flux [209]. Thus, aging studies strongly imply a relationship between proteostasis and nutrient sensing.

3.2.4   Implications of the actin cytoskeleton in age-related pathologies

Actin and actin-dependent processes that are critical for our bodily defense decline with age. For example, phagocytosis of microorganisms by macrophages relies on the activation of the Arp2/3 complex and the resulting increase in actin polymerization and branched network formation. During bacterial infection, alveolar macrophages activate Arp2/3 mediated filipodia network through a Rho GTPase, Rac1 [210]. However, aged mice exhibit a reduction in Rac1 expression, which raises the possibility that the increased susceptibility to sepsis following respiratory infection in elderly subjects may be due to age-linked declines in the actin cytoskeletal function in phagocytosis [210].

Similarly, there are age-linked declines in 1) migration of mesenchymal stem cells (MSCs) to sites of injury and regeneration in bone, skin, liver, and muscle, and 2) cell surface shape changes and recruitment of receptors to the site of interaction between T-cells and antigen-presenting cells (APCs) during the formation of the immunological synapse [211–216]. These processes are dependent on the actin cytoskeletal dynamics, which declines with age [211–216]. Indeed, aged CD4$^+$ T-cells display a significant decline in F-actin accumulation at the interface [212, 213, 215]. Moreover, stimulated aged CD4$^+$ T-cells studies shows only 20% of these cells had F-actin accumulation at the interface compared to 66% in the younger cells [214]. Thus, defects in actin dynamics with age may contribute to age-linked declines in tissue regeneration and the immune response.

Other studies support a role for age-associated declines in stability of actin filaments in actin organization and function in *C. elegans*. The integrity of the actin cytoskeleton declines in the hypodermis, muscles, and intestines in *C. elegans* with advanced age [217]. The observed change in cytoskeletal integrity is dependent on *hsf-1*, a transcription factor that regulates expression of heat shock proteins and *pat-10*, a protein that affects F-actin stability [217, 218]. Deletion of *hsf-1* or *pat-10* results in premature aging of the actin cytoskeleton as well as reduced stress resistance and lifespan. Conversely, overexpression of *hsf-1* or *pat-10* results in tissue-specific protection of actin organization as well as increased actin stability, thermotolerance and lifespan. Finally, expression of *hsf-1* declines with age. Together, these studies indicate that age-linked declines *hsf-1* reduces lifespan through effects on *pat-10* expression and the associated effects on actin stability in *C. elegans*.

My thesis research focuses on actin cables, bundles of F-actin, that are critical for 1) establishment and maintenance of cell polarity, 2) cell division, and 3) asymmetric inheritance of aging determinants during cell division in yeast. My studies support a role for the stability of actin filaments within actin cables (Chapter 2) and for bundling of F-actin into actin cables (Chapter 3) in age-linked declines in the yeast actin cytoskeleton. Since declines in actin stability, dynamics and assembly also occur in *C. elegans* and mouse models, it is possible that the mechanisms identified in my studies may extend our understanding of the aging process and the impact of the actin cytoskeleton on that process in other eukaryotes.

4   Figures

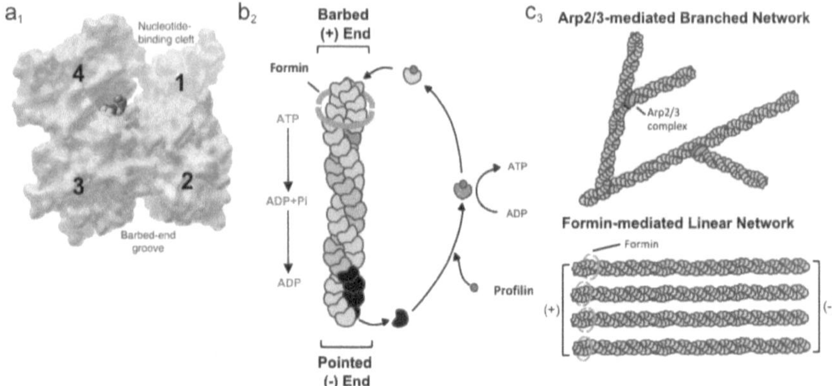

[1] Modified from Pollard Cold Spring Harb Perspect Biol. 2016 Aug; 8(8): a018226 doi: 10.1101/cshperspect.a018226
[2] Modified from Moose and Goode Microbiol Mol Biol. 2006 Sep; 70(3) 605-645 doi: 10.1128/MMBR.00013-06
[3] Modified from Suarez and Kovar Nat Rev Mol Cell Biol. 2016 Sep 14; 17(12) 799-810 doi: 10.1038/nrm.2016.106

Figure 1.1: Schematic of actin polymerization: monomeric G-actin, filamentous actin, and actin networks.

(a) 3D reconstruction of a crystalized monomeric G-actin noted with domains, nucleotide binding cleft, ATP binding site, and directionality of actin, barbed-end groove. (b) Polymerization of actin into filamentous actin. Directionality is labeled in the orientation pointed (-) end to barbed (+) end. Actin turnover is labeled by the hydrolysis of ATP. Regeneration of ATP-bound monomeric G-actin for addition to the barbed (+) end, mediated by actin-binding protein, profilin. (c) Actin filaments are formed into two major networks delineated by actin nucleating promoting factors during polymerization, linear network mediated by formins and branched network mediated by the Arp2/3 complex.

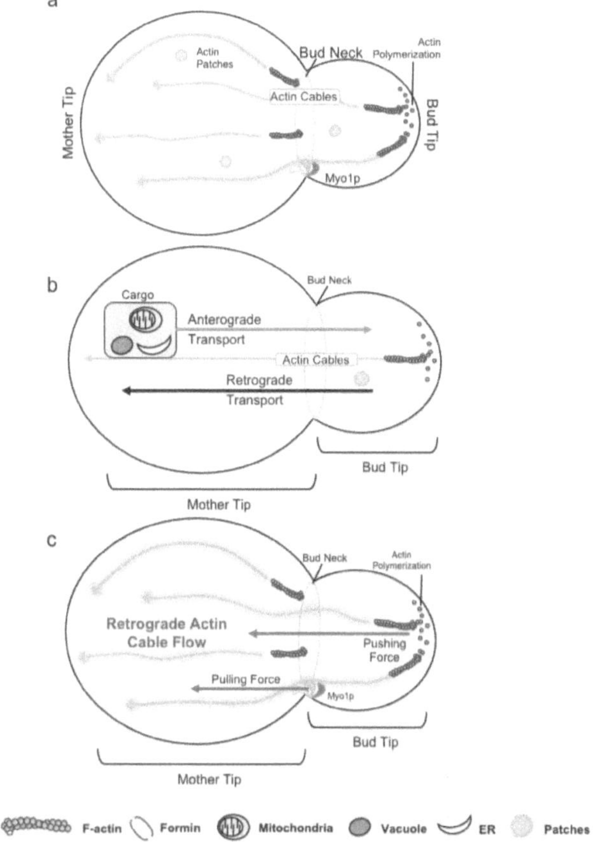

Figure 1.2: Schematic of the actin cytoskeleton assembly, actin-based cargo transport, and actin dynamics in yeast.

(a) Budding yeast actin cytoskeleton forms into major components that persist through the entire cell division: actin cables (linear F-actin bundles) and actin patches (endocytic vesicles coated in branched arp2/3 actin network). Polymerization of actin occurs are two sites: bud tip and bud neck. Elongation of actin cables occur from the bud tip or bud neck to the mother cell tip. (b) Budding yeast cargo is transported along actin cables between mother and daughter cell. Cargo such as: mitochondria, secretory vesicles, vacuole, and ER are transported in the anterograde direction, mother to daughter cell. In contrast, actin patches are transported in retrograde direction, daughter to mother cell. (c) Similar to mammals, budding yeast actin cytoskeleton is dynamic and undergoes a retrograde actin cable flow (RACF) where cables are continuously formed in daughter cells and elongated to the mother cells. RACF is generated by two forces, pushing forces (actin polymerization) and pulling forces (type II myosin protein, Myo1p).

29

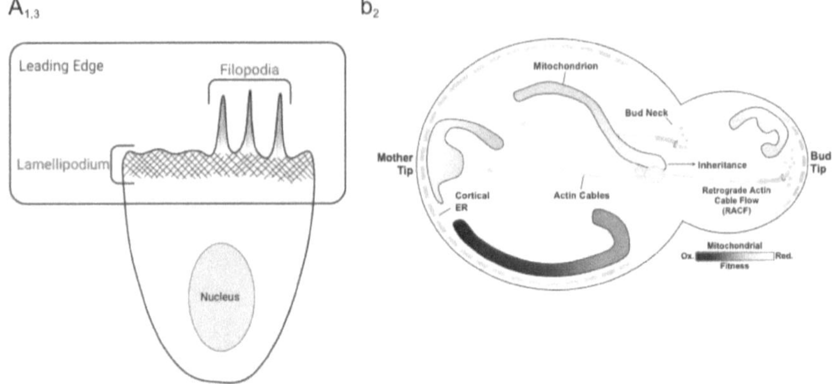

¹ Modified from Suarez and Kovar Nat Rev Mol Cell Biol. 2016 Sep 14; 17(12) 799-810 doi: 10.1038/nrm.2016.106
² Higuchi et al. Curr Biol. 2013 Dec 02; 23(23) 2417-2422 doi: 10.1016;j.cub.2013.10.022
³ Created with BioRender.com

Figure 1.3: Schematic of actin dynamic function in mammals and budding yeast.
(**a**) Actin turnover, dynamics, is responsible for forming the filopodium, lamellipodium in a cell. These two cellular processes generate a protrusive mechanical force enabling a whole-cell migration. (**b**) Retrograde actin flow *Saccharomyces cerevisiae* is conserved in yeast and known as the retrograde actin cable flow, the directional polymerization actin and formation of actin cables occurs at the bud tip and bud neck then extended throughout the cell to the mother distal tip. Concurrently, mitochondria are trafficked along actin cables in the anterograde direction, mother to bud.

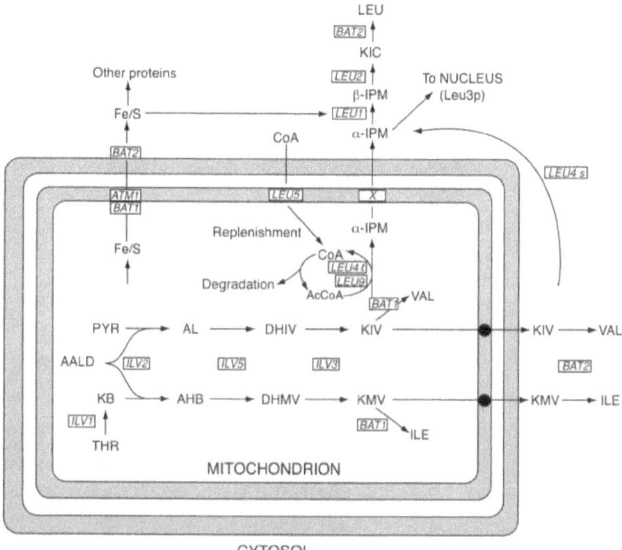

Kohlhaw Microbiol Mol Biol Rev. 2003 Mar;67(1):1-15 doi: 10.1128/mmbr.67.1.1-15

Fig. 1.4: Schematic of BCAA synthesis at the mitochondrion in *Saccharomyces cerevisiae*.
The synthesis of branched-chain amino acids at the mitochondria. BCAA synthesis begins with catalyzing either 2 pyruvates or threonine intermediate, ketobutyrate, to form 2-acetolactate (valine and leucine precursor). The α-ketointermediate, KIV, in the valine pathways also acts as a leucine precursor, marking off a branching point to synthesize leucine, which include 3 additional steps. All BCAAs can either be synthesized in the mitochondria or outside the mitochondria depending which paralog branched-chain amino acid transaminase protein, Bat1p or Batp2, is used.

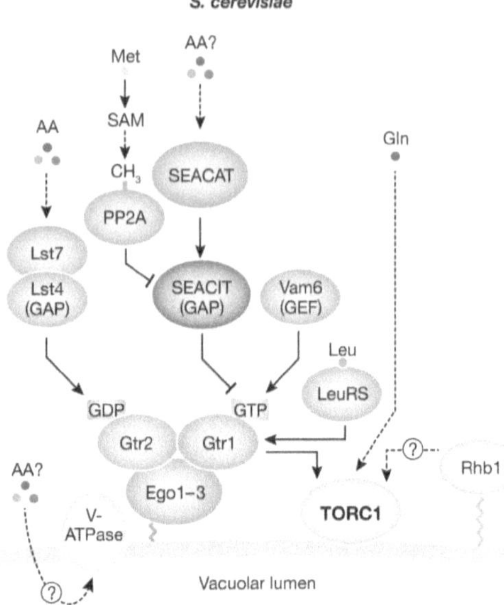

Gonzalez et al. EMBO J. 2017 Feb 15; 36:397-408 doi: 10.15252/embj.201696010

Figure 1.5: Schematic of TORC1 and its upstream regulators in *Saccharomyces cerevisiae*.
A master regulator, TORC1, coordinates the environment nutritional status to anabolic processes is highly regulated. TORC1 is constitutively associated to the vacuolar membrane, where its activity is regulated by the EGO complex (Ego1-3; GTPases Gtr1/2p). Active forms of GTPases that lead to TORC1 activity are: Gtr1p-GTP and Gtr2p-GDP. The GTPases-activating protein (GAP) for Gtr1p hydrolysis is regulated by a complex called SEACIT, which is further regulated by SEACAT. Gtr1p is further recycled by its GEF protein, Vam6p. Gtr2p GAP protein complex is Lst4/7p, however, its GEF protein is currently unknown. Currently, the literature has discovered 3 amino acid sensors that sense amino acids and regulate TORC1 activity: Lst4/7p, SEACAT, and LeuRS.

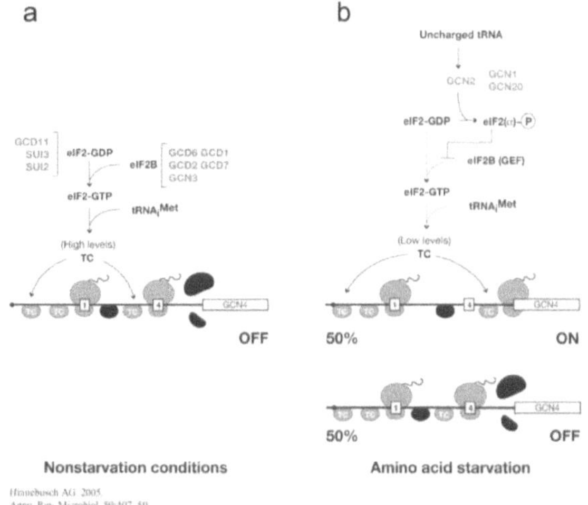

Hinnebusch Annu Rev Microbiol. 2005 59:407-50 doi: 10.1146/annurev.micro.59.031805.133833

Figure. 1.6: Schematic of depression of transcription factor Gcn4p during activated GAAC pathway.
(a) Under amino acid rich conditions, the general amino acid control (GAAC) pathway is repressed and translation is initiated where active tRNA,Met bound eIF2 - GTP binds to 40S ribosome forming a 43S complex (ternary complex - TC), which scans for the corresponding base pair start codon and binds to 60S ribosome to complex the formation of 80S ribosome to initiate translation. Additionally, recycling of eIF2 to its GTP state occurs quickly increasing the formation of TC concentration and at high concentration 40S subunit quickly rebinds a new TC in time to reinitiate at μORF4 and because μORF is the strongest inhibitory translation barrier it dissociates of the ribosome complex before Gcn4 mRNA (b) Under amino acid limiting conditions, GAAC is activated by Gcn2p recognizing the accumulation of uncharged tRNAs triggering its kinase activity to phosphorylate eIF2α to inhibit protein translation, however at the same time, paradoxically promotes the translation of Gcn4 mRNA. Gcn4 mRNA contains 4 additional upstream short ORFs (μORF) that act as translation barriers: μORF1 a "leaky" barrier and μORF4 is a strong translation barrier. Under amino acid starvation conditions where TC formation is low due to the decrease recycling of eIF2 to its GTP bound state, half of the rescanning 40S ribosome fail to rebind TC until scanning past μORF and reinitiate at the Gcn4 mRNA instead, while the remaining half can bind to TC before μORF4 and initiate translation of μORF4 then dissociate before Gcn4 mRNA.

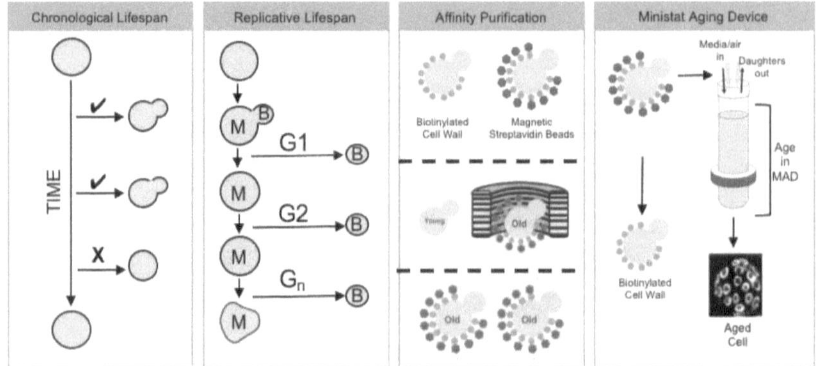

[1] Modified from Lippuner et al. FEMS Microbiol Rev. 2014 Mar;38(2):300-25. doi: 10.1111/1574-6976.12060
[2] Modified from Hendrickson et al. eLife. 2018 7:e39911 doi: 10.7554/eLife.39911

Figure 1.7: Methods to track the age of a yeast lifespan.
(a) Chronological lifespan measures how long a cell can survive in a non-dividing state known as stationary phase. (b) Replicative lifespan is determined by tracking the number buds emerged from a mother cell before she senescence. (c) Affinity based aging cell enrichment, which isolates large batches of young and old cells through a single-timepoint labeling with biotin on cell walls and can repeated with subsequent rounds followed by affinity purification with streptavidin magnetic beads. (d) Modified aging enrichment program, Ministat, which removes the a single-timepoint labeling and provides cells with a consistent flow of fresh media to allow continuous growth for up to 72 hours, rather than conducting multiple rounds to enrich for an older population than the first.

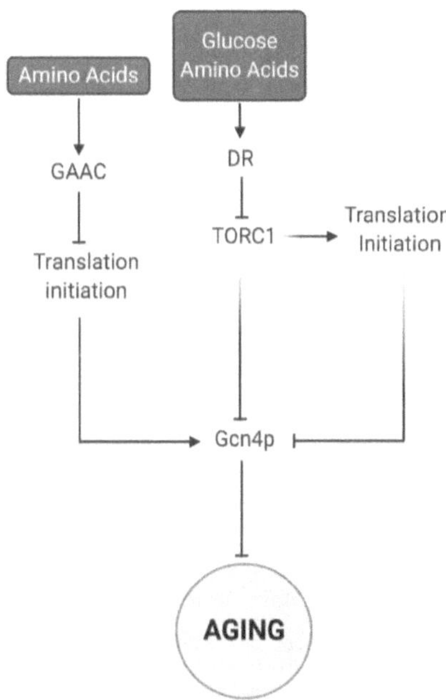

¹ Modified from Steffen et al. Cell. 2008 133(2):292-302. doi: 10.1016/j.cell.2008.02.037
² Created with BioRender.com

Figure 1.8: Schematic of major nutrient sensing pathways in lifespan control.
Two major nutrient sensing pathways that regulate lifespan are TORC1 and GAAC. TORC1 can sense both carbon and nitrogen source changes (diet restriction), while GAAC can only sense nitrogen source changes. When the nitrogen source is low, amino acid limiting conditions: 1) can activate the GAAC pathway to inhibit global synthesis to shift its energy to translate Gcn4p or 2) inhibit TORC1 activity and inhibiting its downstream pathways, including Gcn4p translation. Regulating Gcn4 translation was found to be key to controlling the replicative lifespan in yeast.

Chapter 2: A novel aging determinant regulates the actin cytoskeleton, nutrient sensing, and lifespan in yeast.

1   Summary

In yeast, actin cables are F-actin bundles that are essential for cell division through their function as tracks for cargo movement from mother to daughter cell. Actin cables also affect yeast lifespan by promoting transport and inheritance of higher-functioning mitochondria to daughter cells. Here, we report that actin stability declines with age and, conversely, stabilization of actin promotes mitochondrial function and healthspan. Our genome-wide screen for genes that affect actin cable stability led to the open reading frame, *YKL075C*. Deletion of *YKL075C* results increased actin cable stability and abundance, mitochondrial fitness, and replicative lifespan. Transcriptome analysis revealed a role for *YKL075C* in regulating branched-chain amino acids (BCAAs) metabolism. Consistent with this, modulation of BCAAs metabolism or decreasing leucine levels promotes actin cable stability and function in mitochondrial quality control. Our studies support a role for actin stability in lifespan control in yeast, and for a novel role *YKL075C* and BCAAs in that process.

2   Introduction

The actin cytoskeleton is essential for cellular homeostasis, serving not only as the cell's architectural support but also as the source of contractile and propulsive forces for cell motility and cytokinesis, intracellular organelle movement, phagocytosis, and endocytosis as well as inter- and intracellular communication [1, 3]. Recent studies also support a role for the actin cytoskeleton in lifespan control. Specifically, age-associated declines in the actin cytoskeleton compromise the immune system by inhibiting 1) phagocytosis and reactive oxygen species (ROS) production by macrophages [210], 2) ROS production and exocytosis in neutrophils [219–223], 3) migration of mesenchymal stem cells to sites of injury [216], and 4) formation of the immunological synapse between T-cells and antigen-presenting cells [211–214, 220, 224]. Similarly, the speed of actin sliding on skeletal muscle myosin and maximum velocity of muscle fiber shortening decline with advanced age, which may contribute to age-linked loss of skeletal muscle function [225–229].

Studies in *Caenorhabditis elegans* revealed a mechanism for loss of actin function with age that relies on HSF-1, a transcription factor that regulates the expression of heat shock genes and of PAT-10, a calcium-binding actin-associated protein that affects F-actin stability [217, 218]. Expression of *hsf-1* declines with age in a manner that resembles age-associated declines in the integrity of the actin cytoskeleton in muscle, intestine, and hypodermis. Moreover, deletion of *hsf-1* or *pat-10* results in premature aging of the actin cytoskeleton as well as reduced stress resistance and lifespan. Conversely, overexpression of *hsf-1* or *pat-10* results in tissue-specific protection of actin organization as well as increased actin stability, thermotolerance, and lifespan. Together, these studies indicate that age-linked declines in F-actin stability affect cellular fitness during stress and lifespan control, and a role for HSF-1 and PAT-10 in actin stability during aging in *C. elegans*.

In budding yeast and human mammary stem-like cells, mitochondria are differentially partitioned during asymmetric cell division, which in turn, affects daughter cell fate [71, 72]. In yeast, this process affects lifespan and is dependent upon actin cables, actin bundles that serve as tracks for transport of mitochondria and other cargos from mother to daughter cell. Actin cables are also dynamic structures that undergo retrograde flow, a conserved mechanism for assembly- and motor-driven movement of actin cables from their site of assembly in the bud into the mother cell [57]. As a result, mitochondria are effectively "swimming upstream" as they move from mother to daughter cells on dynamic actin cables. Indeed, increasing the rate of retrograde actin cable flow (RACF) promotes inheritance of fitter mitochondria to daughter cells, extends lifespan and promotes healthspan (the period of life spent in good health, free from disease or disabilities associated with age) [70]. These findings support the model that actin dynamics and, more specifically, RACF affects cell fitness and lifespan by serving as a filter to promote movement and preferential inheritance of higher functioning mitochondria from mother to daughter cells during asymmetric yeast cell division. Since mitochondria are differentially inherited in stem-like cells, it is possible that mechanisms for mitochondrial quality control during asymmetric cell division that are identified in yeast function in other eukaryotes.

Here, we obtained evidence that actin cable stability declines with age, and promoting actin stability promotes mitochondrial function and cellular healthspan using yeast as a model system. We also identified a role for a previously uncharacterized gene (*YKL075C*) in actin cable stability, mitochondrial quality control, and lifespan. Finally, we find that *YKL075C* affects actin cables through TORC1-independent effects on branched-chain amino acid (BCAA) metabolism.

3   Results

3.1   Actin cables undergo a decline in stability with age in yeast

We tested whether actin cable organization changes with age in yeast, as in *C. elegans*. Actin cables and patches are F-actin containing structures that persist throughout the yeast cell division cycles. As described above, actin cables are essential for yeast cell division and serve as tracks for the movement of cellular constituents from mother cells to buds and for asymmetric inheritance of mitochondria. They are resolved as tubular structures that align along the mother-bud axis and extend from the bud tip or neck to the tip of the mother cell distal to the bud in mid-log phase yeast cells. Actin patches, F-actin-coated endosomes, are formed and localize to the bud during early stages of the cell cycle and to the bud neck late in the cell cycle [20].

The model that we use for yeast cell aging is replicative lifespan (RLS, the number of times a mother cell divides before senescence), a model for aging in cell division-competent cells [153]. Here, old cells were isolated using a modification of a method, whereby mid-log phase cells, which are primarily young cells, are labeled with biotin, propagated to specific replicative ages, and isolated by streptavidin affinity chromatography [157]. The old cells used in these studies have undergone 2 rounds of replication, as assessed by quantitation of bud scars, ring-shaped structures that are deposited at sites of daughter cell formation on the mother cell wall. Thus, the old cells used in these studies had undergone approximately 50% of their total replicative lifespan.

Using structured illumination super-resolution microscopy (SIM), we detect on average 8-actin cables with an apparent diameter of 0.2 µm in the mother cell of young yeast cells (Fig. 2.1a-c). Moreover, we observe a significant decrease in the apparent diameter of actin cables and an increase in actin cable number in old compared to young yeast cells (Fig. 2.1a-c). Thus, we find that age-associated changes in the organization of the actin cytoskeleton occur in budding yeast.

Next, we tested whether actin cables destabilize with age in yeast by analyzing the sensitivity of actin cables to destabilization by Latrunculin A, a small molecule that promotes F-actin depolymerization by binding to, sequestering and inhibiting polymerization of monomer G-actin. Treatment of yeast with high levels of Lat-A (above 10 µM) results in loss of all components of the yeast actin cytoskeleton. However, exposure to low concentrations (1-10µM) of Lat-A results in preferential loss of actin cable [230]. The increased sensitivity of actin cables to Lat-A indicates that F-actin within those structures are more dynamic compared to F-actin within actin patches. Therefore, we tested the effect of low levels of Lat-A on actin cable stability as a function of replicative age in yeast.

For these studies, old cells were isolated using a miniature-chemostat aging device (mini-stat) [158]. Mid-log phase cells, which are primarily young cells, were labeled with biotin conjugated to magnetic beads and introduced into the mini-stat. A magnetic field was then applied across the mini-stat, to retain biotinylated cells, and progeny from those cells, which are not biotinylated, were eluted from the device using a constant feed of fresh media. Old cells, which aged during propagation in the mini-stat for 45 hrs, were then harvested from the device by removal of the magnetic field and elution with fresh media. The average age of old cells used in these studies is 24 generations.

To assess the effects of advanced age on actin cable stability, young and old yeast cells were treated with 1µM Lat-A and the actin cytoskeleton within those cells was visualized as a function of the time of Lat-A treatment. We observed a loss of actin cables with increasing time of Lat-A treatment in young cells. However, our analysis of old cells revealed a more rapid rate of loss of actin cables with Lat-A treatment (Fig. 1d,e). Thus, actin cables undergo age-linked declines in stability in budding yeast as in *C. elegans*. This, in turn, could contribute to the decrease in actin cable width and increase in actin cable abundance in yeast at advanced age.

3.2 Stabilization of the actin cytoskeleton affects cellular healthspan and lifespan

We used the *act1-159* mutation to test the effects of actin stabilization on yeast cell fitness, replicative lifespan, and healthspan. The *act1-159* mutation is a single amino acid substitution at position

159 (V159N) in the actin encoding *ACT1* gene, which inhibits conformation changes in actin protein during Pi release and reduces the rate of F-actin depolymerization; thus delaying actin turnover producing more stable actin cables actin stability [231]. Additional to its actin stability characteristics, yeast expressing the *act1-159* are known to exhibit defects in the actin polymerization/depolymerization cycle and endocytosis; an increase in actin cable length, and a decrease in loss of actin cables in response to mutations in actin cable stabilizing proteins (fimbrin, Mdm20 and tropomyosin) [231].

We used CRISPR technology to introduce the V159N substitution into the *ACT1* gene, the sole actin gene of yeast, at its chromosomal locus. We confirmed that the mutation results reduced sensitivity to actin cable destabilization by treatment with low levels of Lat-A. (Fig. 2.1f). Importantly, we found the *act1-159* mutation does not affect the rate of RACF. Thus, the mutation affects F-actin and actin stability but has no obvious effect on the rate of actin cable assembly (Supplementary Fig. 2.1a,b).

We tested the effect of the *act1-159* mutation on mitochondrial function using a ratiometric sensor for the redox state of the mitochondrial matrix (mito-roGFP1) [71, 232, 233]. Using mito-roGFP1, we found that mitochondria are asymmetrically inherited in *act1-159* cells: mitochondria in buds are more reduced and, therefore, higher functioning compared to mitochondria in mother cells in *act1-159* cells, like wild-type cells (Supplementary Fig. 2.1c). Equally important, we find that mitochondria are more reduced in *act1-159* mutants compared to mitochondria in wild-type cells (Fig. 2.1g,h). Thus, increasing actin stability is sufficient to promote mitochondria function. Since actin cables drives inheritance of fitter mitochondria to yeast daughter cells, the increase in mitochondrial function observed in the *act1-159* mutant may be due to the increase in actin cable stability.

Unexpectedly, we found that the *act-1-159* mutation has complex effects on lifespan and healthspan. We detect a decline in the replicative lifespan in *act1-159* mutants compared to wild-type cells (Fig. 2.1i). This contradicts our knowledge of improved mitochondria quality extends replicative lifespan [70]. However, a consequence to increase actin stability in *act1-159* cells exhibit defects in actin polymerization and depolymerization, which considering the critical importance of actin dynamics during cell division, it is possible that increased stability may have a detrimental effect on cell division, and thus inversely correlate with replicative lifespan. However, healthspan was improved when increase actin stabilization. To evaluate healthspan, we used mean generation time (MGT) or the time to complete a single division cycle [150]. Under our experimental conditions, the mean generation time ranges from 100 - 150 min for young wild-types cells and increases to 200-300 min as yeast cells age. Consistent with previous findings, the healthspan of *act1-159* cells early is greater to that of age-matched wild-type cells (Fig. 2.1j). Therefore, the reduced lifespan and healthspan observed upon stabilization of F-actin early in replicative life may be due to detrimental effects on mitochondria-independent processes.

In contrast, we found that MGT at a late age (past replicative age 20) tends becomes more reduced in *act1-159* cells while wild-type cells generally exhibit a significant increase in mean generation due to cellular dysfunction at a late age (Fig. 2.1j). Thus, increasing F-actin stability can have a positive effect on healthspan late in the replicative lifespan of yeast, potentially by compensating for age-linked destabilization of the actin cytoskeleton. Overall, our data suggest that controlling actin stability may be key to delay the age-associated onset of loss of integrity and function of the actin cytoskeleton.

3.3   Discovering a novel actin regulatory factor that regulates longevity

Our data clearly show that there is a measurable decrease in actin stability with age and F-actin stabilization can promote health at a late age. However, it is not clear whether the effects are due to effects on the integrity of actin cables, as opposed to stabilizing all F-actin containing structures. Therefore, we carried out a genome-wide screen whose deletion specifically affect actin cable stability, henceforth, a novel actin cable regulator. First, we confirmed previous findings that actin cables are highly sensitive to Lat-A and treatment with low levels of Lat-A (<10µM) results in selective loss of actin cables and severe inhibition of yeast cell growth, but has no obvious effect on morphology, localization or polarization of actin patches (Fig. 2.2a) [230].

Therefore, we screened for yeast genes, which when deleted, reduced the sensitivity of yeast to the growth-inhibiting effects of treatment with 10 µM of Lat-A (Supplemental Fig. 2.2a). 18 deletion strains exhibited growth rates in the presence of low levels of Lat-A that are greater than those observed in wild-type cells (Supplemental Fig. 2.2a,b). We detected no obvious defects in actin polarity, however, 13 of the 18 strains analyzed exhibit a significant increase in actin cable number and no strains contain fewer actin cables than wild-type cells (Fig. 2.2b). Although the majority of the deletion mutants exhibited more actin cables, there was no correlation between the growth rate in the presence of Lat-A and actin cable content.

One of the strongest hits from our screen is an uncharacterized open reading frame, *YKL075C*. Deletion of *YKL075C* does not affect yeast cell growth: the growth rate of the *YKL075C* null strain from the yeast deletion collection and newly generated *YKL075C* deletion strains are similar to that of wild-type cells. Interestingly, the growth rate of all *YKL075C* null strains in the presence of Lat-A is significantly greater than Lat-A treated wild-type cells (Fig. 2.2c). Thus, deletion of *YKL075C* decreases the sensitivity to the growth-inhibiting effects of Lat-A.

To determine whether the reduced sensitivity of *YKL075C* null strains to Lat-A is suggestive of increased actin stability, thus should be reflected in the actin cable morphology. We visualized those structures in living wild-type and *YKL075C* nulls strains. We find that deletion of *YKL075C* is sufficient to increase actin cable abundance (Fig. 2.2d,e). Next, we tested the sensitivity of actin cables in *YKL075C* null cells to exposure to low levels of Lat-A. We found that *ykl075c*Δ cells had a significantly reduced rate of actin cable loss under Lat-A exposure compared to wild-type cells (Fig. 2.2f,g). Thus, the reduced sensitivity of *ykl075c*Δ cells to the growth-inhibiting effects of Lat-A may be due to increased stability and reduced Lat-A sensitivity of actin cables in *YKL075C* nulls strains.

Finally, we found that deletion of *YKL075C* has profound effects on mitochondrial quality control, lifespan, and cellular fitness during aging. Using mito-roGFP1, we find that mitochondria are more reduced and therefore higher functioning in *ykl075c*Δ compared to wild-type cells (Fig. 2h, i). Additionally, *ykl075c*Δ cells exhibited asymmetric inheritance of mitochondrial quality (Supplementary Fig. 2c). Interestingly, no detectable difference in RACF velocity in *ykl075c*Δ compared to wild-type cells (Supplementary Fig. 2.2d,e). Thus, the increased mitochondrial function observed in *ykl075c*Δ cells may be due to increased stability of actin cable and not to effects on actin cable assembly.

Equally important, we find that deletion of *YKL075C* results in a significant increase in RLS (Fig. 2j). The mean RLS of wild-type cells is 19 generations. In contrast, the mean RLS of *ykl075c*Δ cells are 26 generations. Consistent with this, we find that deletion of *YKL075C* results in an increase in healthspan, as assessed by quantitation of MGT as a function of age (Fig. 2.2k). Thus, we identified a novel role for the previously uncharacterized protein, Ykl075cp, in regulating actin cable stability and abundance, mitochondrial quality control, RLS, and organismal healthspan.

3.4   *YKL075C* targets branched-chain amino acid homeostasis to regulate the actin

cytoskeleton

YKL075C is an uncharacterized open reading frame with no obvious functional domains or sequence similarity to known proteins. Therefore, we used unbiased approaches to study its properties and function. First, we tagged the endogenous *YKL075C* gene at its C-terminus with 13 copies of the Myc epitope and tested whether the gene encodes a protein and if so, the localization of that protein. Using western blot analysis, we found that the *YKL075C* gene encodes a protein with an apparent molecular weight that is expected based on the predicted amino acid composition of the protein (Supplementary Fig. 2.3a). Using immunofluorescence, we found the tagged protein localizes to punctate cytoplasmic structures that are evident in the mother cells and buds (Fig. 2.3a).

Next, we performed RNA-seq analysis to determine the effect of deletion of *YKL075C* on gene expression in mid-log phase yeast (Table 2.3). Our transcriptome analysis revealed major changes in amino acid, nucleic acid, and protein metabolism as well as changes in macromolecular transport (Fig. 2.3b). Interestingly, GO analysis revealed that 8 of the 10 branched-chain amino acid (BCAA) synthetic genes are differentially expressed in *ykl075cΔ* cells (Fig. 2.3c). Specifically, we find transcriptional changes on key paralog BCAA transaminases (BAT) proteins, Bat1p and Bat2p: downregulation of *BAT1* (preferentially mediates BCAA synthesis) and upregulation of *BAT2* (preferentially mediates BCAA catabolism) [96, 97]. Our qPCR confirmed these opposing transcriptional changes such we find a 50% reduction in *BAT1* expression and a 2-fold increase in *BAT2* expression (Fig. 2.3d,e).

Our results implicate a change in BCAA levels in *ykl075cΔ* cells. We find reduced intracellular levels of free BCAAs in protein lysates of wild type, *ykl075cΔ*, *bat1Δ*, and *ykl075cΔbat1Δ* cells compared to wild-type cells (Fig. 2.3f). Importantly, we observe a more dramatic decline in BCAA levels in *bat1Δ* compared to *ykl075cΔ* cells and no additive effects of the simultaneous deletion of *YKL075C* and *BAT1* on BCAA levels. These findings indicate that 1) *BAT1* is the primary regulator of BCAA levels and 2) *YKL075C* and *BAT1* function in the same pathway for BCAA control.

TORC1 (Target of rapamycin complex 1) is a hallmark nutrient sensor known to coordinate environmental nutrient properties with anabolic cellular processes [102, 234]. Moreover, changes to BCAAs have a well-established effect on TORC1 activity: withdrawing leucine from the growth medium reduces

TORC1 activity and reducing TORC1 activity controls lifespan [112, 119, 235–237]. Therefore, we tested whether Ykl075cp function is dependent on TORC1. Analysis of our transcriptome datasets revealed that the pattern of gene expression *ykl075cΔ* cells are not reflective of a changing TORC1 signaling pathway.

To test whether TORC1 activity differs in *ykl075cΔ* cells, we analyzed responses to rapamycin, a TORC1 inhibitor that causes cell growth arrest [102]. Specifically, we measured the rate of re-entry of cells into the cell cycle after rapamycin treatment [109, 238]. Deletions of *TOR1*, *YKL075C* or double deletion of *TOR1* and *YKL075C* has no effect on yeast growth rate on rich, glucose-based solid media; however, *tor1Δ* cells inhibits recovery from rapamycin treatment: *tor1Δ* cells exhibit reduced growth rates after removal rapamycin from the medium compared to wild-type cells (Fig. 2.3g,h). On the other hand, the recovery of *ykl075cΔ* cells from this treatment was similar to that observed in wild-type cells and the growth of *ykl075cΔtor1Δ* double mutants was similar to that of *tor1Δ* mutants (Fig. 2.3g,h). These results indicate Ykl075cp function may be independent of the TORC1 signaling pathway.

Thus, we directly tested the effects of deleting *YKL075C* on TORC1 activity by examining the phosphorylation of S6 (Rps6p), a TORC1 target. We confirmed previous findings that deletion of *TOR1* reduces Rps6p phosphorylation, while rapamycin treatment fully inhibits TORC1 activity by blocking Rsp6 phosphorylation. Moreover, we find Rsp6p phosphorylation in *ykl075cΔ* cells is indistinguishable from that observed in wild-type cells with no apparent defects in the phosphorylation of Rps6p (Supplementary Fig. 2.3b,c). Taken together, we find that deletion of *YKL075C* does not alter TORC1 activity in yeast. Thus, the cytoskeletal regulation through changes to BCAA metabolism observed in *ykl075cΔ* cells is independent of TORC1.

3.5 Actin cytoskeleton dependent longevity is regulated by BCAA levels

Here, we tested whether changes in BCAA levels can drive changes in cytoskeletal integrity by manipulating *BAT1* or *BAT2* expression or BCAA levels in the growth-medium of wild-type and *ykl075cΔ* cells. Although deletion of *BAT2* has no obvious effects on actin cables (Supplementary Fig. 2.4a,b), *bat1Δ* or *ykl075cΔ* mutants exhibit an increase in actin cable abundance compared to wild-type cells (Fig. 2.4a,b). Consistent with this, depletion of leucine for 20 min is sufficient to increase actin cable abundance of wild-type cells and leucine supplementation results in reduced F-actin phalloidin signal intensity, indicating reduced F-actin content, as well as defects in the morphology and polarity of actin cables in both wild-type and *ykl075cΔ* cells (Fig. 2.4c,d; Supplementary Fig. 2.4c,d). Taken together, these findings indicate that reducing BCAA, and more specifically, reducing leucine levels is sufficient to drive an increase in actin cable abundance. Finally, we find that simultaneous deletion of *BAT1* and *YKL075C* or depletion of leucine in *ykl075cΔ* did not have additive effects. These findings indicate that *YKL075C* and BCAA function in the same pathway for control of actin cable abundance.

To determine whether BCAA affects actin cable stability, we exposed *bat1Δ* cells to low concentrations of Lat-A for 20-minutes and conducted time-course analysis of actin cables as described above. We found the rate of actin cable loss in *bat1Δ* cells is comparable to *ykl075cΔ* cells and significantly lower than wild-type cells (Fig. 2.4e,f). Moreover, Lat-A treatment to either *ykl075cΔ* and *bat1Δ* cells and does not have an additive effect on the rate of cable loss (Fig. 2.4e,f). These findings indicate that the increased actin cable abundance observed in *bat1Δ* cells is due to effects on actin cable stability.

Finally, we measured the effect of modulating BCAA metabolism on mitochondrial function using mito-roGFP1. First, the asymmetric inheritance of mitochondria is not affected in *bat1Δ* cells (Supplementary Fig. 2.4f). On the other hand, mitochondria in *bat1Δ* cells, similar to *ykl075cΔ* cells, are more reduced and therefore higher functioning compared to mitochondria in wild-type cells. Moreover, mitochondrial redox state in *bat1Δ* and *ykl075cΔ* cells are indistinguishable (Fig. 2.4g,h). Thus, reducing BCAA biosynthesis and *YKL075C* promote mitochondrial function through similar pathways.

4   Discussion

No organism can evade the decline of cellular functions, including actin cytoskeletal function, produced by the aging process. A decline in actin dynamics, which results in defects in cell migration, shape changes, and signal transduction, occurs in mammalian cells [214, 216]. Recent studies in *C. elegans* indicate that changes in actin stability may underlie age-linked declines in actin function [217, 218]. Here, we provide the first evidence that actin cables undergo age-linked declines in stability in yeast. We observe an increase in actin cable abundance, a decrease in the apparent width of actin cables, and increased sensitivity of actin cable destabilization by low level Lat-A treatment in yeast cells as they age. Our finding that an essential component of the yeast actin cytoskeleton undergoes age-linked declines in stability in yeast, as in *C. elegans*, indicates that observed loss of actin cytoskeletal integrity and function during aging is due to declines in the stability of actin cytoskeleton.

We also find that stabilization of total cellular F-actin has complex effects on lifespan and healthspan in yeast. Specifically, we find that the actin stable mutant, *act1-159*, exhibits an increase in mitochondrial fitness but decreased healthspan and survival early in replicative lifespan. On the other hand, we observe a decline in actin cable stability with age and find that increasing the F-actin stability may promote healthspan during late stages of replicative lifespan. These findings indicate that 1) actin cytoskeletal dynamics and stability are optimized for function in young cells, 2) F-actin stabilization in young cells compromises cell fitness independent of its effect on mitochondrial quality control, and 3) stabilization of F-actin may compensate for the declining integrity and function of the actin cytoskeleton that occurs in advanced age. Overall, our studies support a role for actin stability in lifespan control in yeast, as in *C. elegans*.

To determine whether stabilization of actin cables, as opposed to total cellular F-actin, affects lifespan, we performed a genome-wide screen for genes whose deletion reduces yeast sensitivity to Lat-A-induced actin cable destabilization. Through these efforts, we identified *YKL075C*, a previously uncharacterized open reading frame, whose deletion resulted in an increase in actin cable stability and abundance. Importantly, deletion of *YKL075C* stabilized actin in a manner that was beneficial to cellular health, as it resulted in increased lifespan and healthspan.

Our transcriptome analysis revealed an unexpected link between *YKL075C* and BCAA metabolism. We find that deletion of *YKL075C* decreases BCAA biosynthetic enzymes expression and BCAA levels. Moreover, decreasing BCAA biosynthesis or depleting a specific BCAA, leucine, increases actin cable abundance and stability, while leucine supplementation has the opposite effect. Finally, we obtained genetic evidence that *YKL075C* and BCAA function the same pathway for control of actin cables and that *YKL075C* function.

Previous studies indicate that BCAAs can affect the actin cytoskeleton. High BCAA levels in Maple Syrup Urine Disease models induces actin rearrangements, which is thought to be through RhoA [239]. Moreover, BCAA restrictive diets were found to increase the lifespan of *Drosophila* and reverse its aging insults [240]. Conversely, BCAAs are known upstream regulators of a central nutrient sensing pathway,

TORC1, whereby dampening levels of BCAA decreases TORC1 activity and controlling TORC1 activity is key for lifespan control [119, 120, 181]. Interestingly, our research shows the lifespan extension in *ykl075cΔ* cells is TORC1-independent. Activation of the general amino acid control (GAAC) pathway is another known nutrient sensing pathway that regulates lifespan control [194, 209]. Therefore, it remains to determine what is the underlying mechanism of Ykl075cp effects on lifespan and healthspan. Overall, our findings reveal a novel role for *YKL075C* in BCAA metabolism by specifically reducing leucine levels to regulate actin cable stability. Since stabilization of F-actin can promote healthspan, our findings raise the possibility that *YKL075C* contributes to aging through effects on actin cable stability.

5   Figures

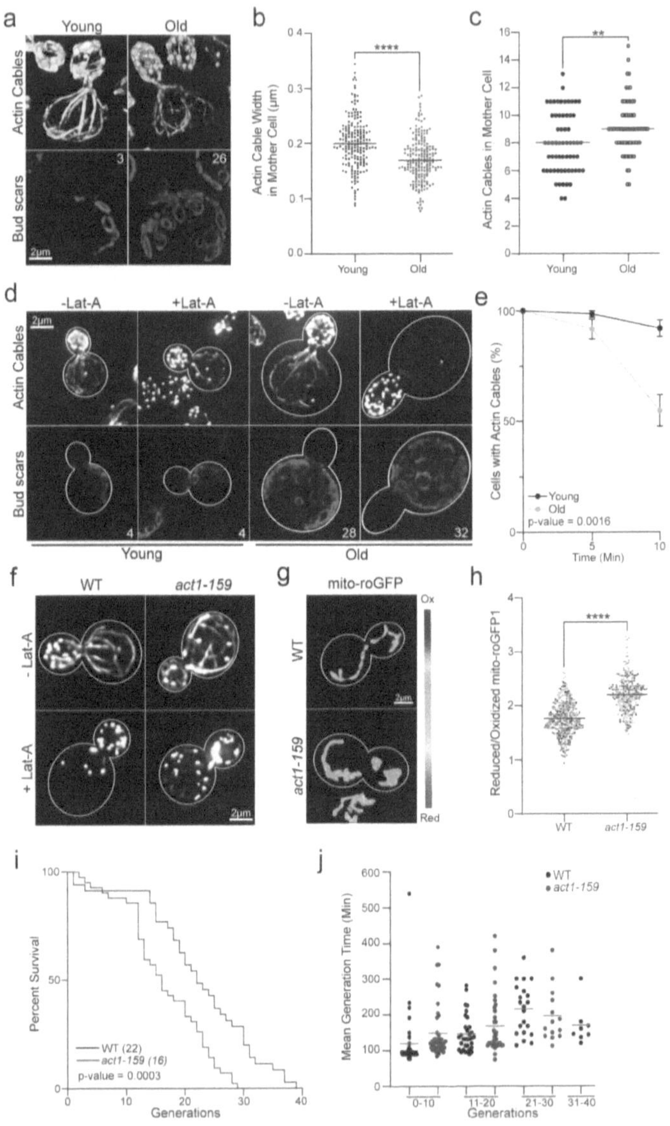
a
Young   Old
Actin Cables
Bud scars
3   26
2µm
b
Actin Cable Width
in Mother Cell (µm)
****
Young   Old
c
Actin Cables in Mother Cell
**
Young   Old
d
-Lat-A   +Lat-A   -Lat-A   +Lat-A
Actin Cables
Bud scars
4   4   28   32
2µm
Young   Old
e
Cells with Actin Cables (%)
Young
Old
p-value = 0.0016
Time (Min)
f
WT   act1-159
- Lat-A
+ Lat-A
2µm
g
mito-roGFP
Ox
WT
act1-159
Red
2µm
h
Reduced/Oxidized mito-roGFP1
****
WT   act1-159
i
Percent Survival
WT (22)
act1-159 (16)
p-value = 0.0003
Generations
j
Mean Generation Time (Min)
WT
act1-159
0-10   11-20   21-30   31-40
Generations

Figure 2.1: Actin stability regulates longevity.

(a) Representative SIM images of AlexaFluor488 phalloidin-stained actin and wheat germ agglutinin 594 (WGA594) stained bud scars in young and old wild-type (WT) cells. (b) Width of actin cables in young and old WT cells isolated from streptavidin columns (n>170 measured points/strain). (c) Actin cable abundance in WT young and old cells isolated from streptavidin affinity columns (n>60 cells/strain). The central bands in panels b-c represents the median. Data are representative from 3 trials. **p <0.01 (non-parametric Mann-Whitney test. ****p <0.0001 (non-parametric Kruskal-Wallis test). (d) Representative SIM images of AlexaFluor488 phalloidin-stained actin and WGA594 stained bud scars in young and old WT cells treated with vehicle (DMSO) and 1μM Latrunculin-A (Lat-A) for 10 min. (e) Sensitivity of actin cables of young and old cells to treatment with LatA. Quantitation of cells that contain actin cables after treatment with Lat-A (1μM) for 0-10 min. Error bars: SEM of 3 trials (n=100 cells/strain/trial). (f) Representative SIM images of actin cytoskeleton in mid-log phase WT or *act1-159* cells that were incubated with vehicle (DMSO) (top panel) or 2μM Lat-A dissolved in DMSO for 20 min (bottom panel). (g) Representative images of reduced:oxidized mito-roGFP1 ratios in WT and *act1-159* cells. Colorimetric scale of redox ratios: higher numbers/warmer colors indicate more-reducing mitochondria. (h) Quantification of reduced:oxidized mito-roGFP1 ratios of WT and *act1-159* cells (n>250 cells/strain from 3 trials). ****p <0.0001, using the non-parametric Mann-Whitney test. (h) RLS of WT vs. *act1-159* cells determined by manual micromanipulation. p-value of 0.0003 (Mantel-cox test). (j) Quantification of mean generation time, time between emerging consecutive buds from the same mother during the lifespan, of all cells as a function of replicative lifespan. Red bar indicates the mean for population. Scale bars in panels a, d, f and g: 2μm.

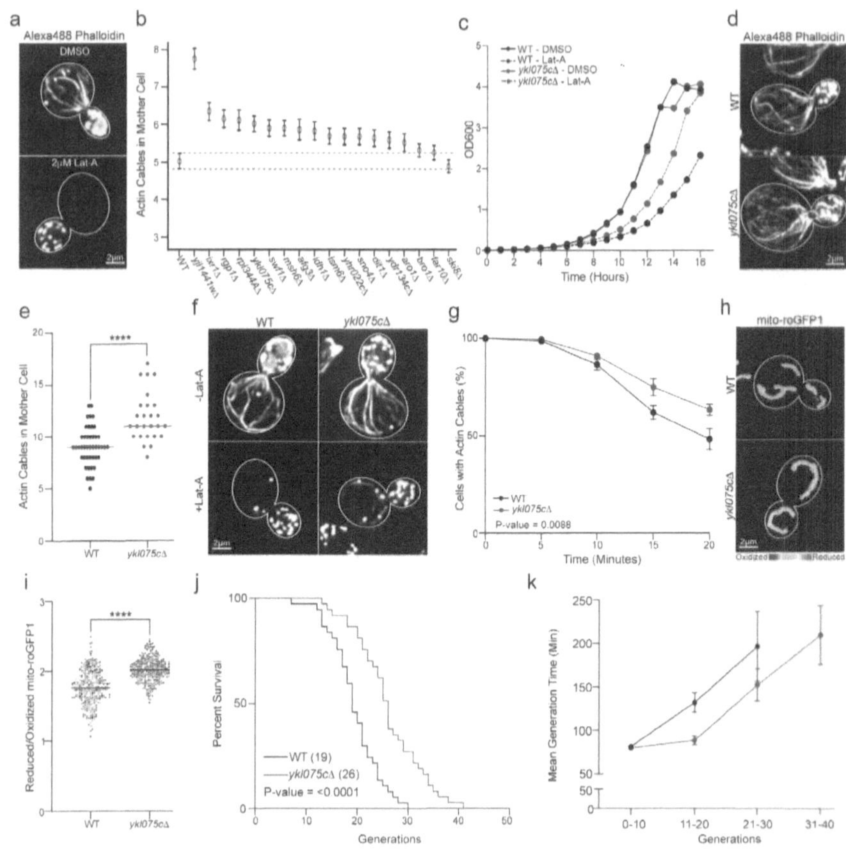

Figure 2.2: Identification of a gene (*YKL075C*) that affects actin cable stability, mitochondria, and lifespan.

(**a**) Representative images of the actin cytoskeleton of WT cells treated with DMSO (vehicle) or 2µM Lat-A for 20 min. Actin was visualized as for Fig. 2.1a. (**b**) 13 hits out of 18 hits form a genome-wide screen against Lat-A show increase in actin cable abundance in comparison to wild-type (**c**) Growth curves of WT and *ykl075cΔ* cells with vehicle (DMSO) or 2µM Lat-A for 16 h (representative from 3 trials). (**d**) Representative images of AlexaFluor488 phalloidin-stained actin cytoskeleton of WT and *ykl075cΔ* cells treated with vehicle (DMSO) (top panel) or 2µM Lat-A for 20 min (bottom panel). (**e**)The presence of actin cable in WT and *ykl075cΔ* cells during 2µM Lat-A treatment. Error bars: SEM of 5 trials (n = 100 cells/strain/trial). p-value = 0.0088 (simple linear regression analysis). (**f**) Representative images of AlexaFluor488 phalloidin-stained actin cytoskeleton of WT and *ykl075cΔ* cells. Scale bar: 2µm. (**g**) Actin cable abundance in mother cells of mid-log phase WT and *ykl075cΔ* cells. The central band represents the median. Data are a representative trial from 3 trials (n>20 cells per strain). ****p <0.0001 (non-parametric Kruskal-Wallis test). (**h**) Representative images of the redox state by mito-roGFP1. Ratio values of reduced:oxidized are presented on a colorimetric scale; higher numbers/warmer colors indicate more-reducing mitochondria. Scale bar: 2µm. (**i**) Reduced:oxidized mito-roGFP1 ratios in WT and *ykl075cΔ* cells (n>250 cells/strain from 3 trials) were determined as for Fig. 2.1g. ****p<0.0001 (non-parametric Mann-Whitney test). (**j**) RLS of WT and *ykl075cΔ* cells determined by manual micromanipulation. p-values, <0.0001 (Mantel-cox test). (**k**) Mean generation time determined as for Fig. 2.1j. Error bars: SEM, n>35 cells/strain). Data presented is a representative trial from 3 trials.

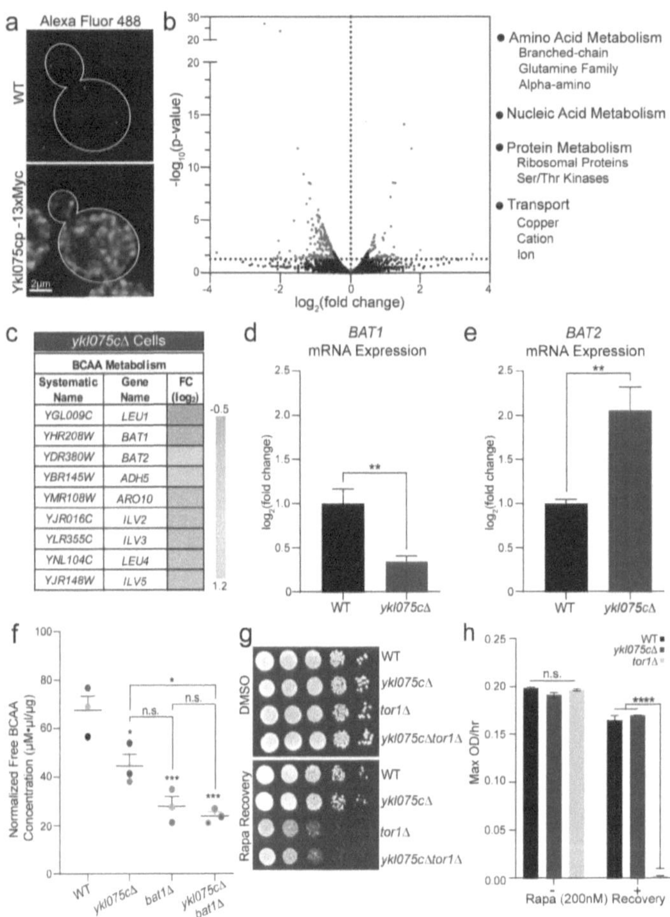
a
Alexa Fluor 488
WT
Ykl075cp -13xMyc
2μm
b
30
25
20
15
10
5
-log10(p-value)
log2(fold change)
-4 -2 0 2 4
Amino Acid Metabolism
Branched-chain
Glutamine Family
Alpha-amino
Nucleic Acid Metabolism
Protein Metabolism
Ribosomal Proteins
Ser/Thr Kinases
Transport
Copper
Cation
Ion
c
ykl075cΔ Cells
BCAA Metabolism
Systematic Name
Gene Name
FC (log2)
YGL009C LEU1
YHR208W BAT1
YDR380W BAT2
YBR145W ADH5
YMR108W ARO10
YJR016C ILV2
YLR355C ILV3
YNL104C LEU4
YJR148W ILV5
-0.5
1.2
d
BAT1
mRNA Expression
log2(fold change)
2.5
2.0
1.5
1.0
0.5
0
WT ykl075cΔ
**
e
BAT2
mRNA Expression
log2(fold change)
2.5
2.0
1.5
1.0
0.5
0
WT ykl075cΔ
**
f
100
80
60
40
20
0
Normalized Free BCAA Concentration (μM•μl/μg)
WT ykl075cΔ bat1Δ ykl075cΔ bat1Δ
n.s.
*
n.s.
***
***
g
DMSO
WT
ykl075cΔ
tor1Δ
ykl075cΔtor1Δ
Rapa Recovery
WT
ykl075cΔ
tor1Δ
ykl075cΔtor1Δ
h
0.25
0.20
0.15
0.10
0.05
0.00
Max OD/hr
WT
ykl075cΔ
tor1Δ
n.s.
****
Rapa (200nM) Recovery
- +

Figure 2.3: Ykl075cp protein localization and links to branched-chain amino acid homeostasis.
(a) Localization of Ykl075cp in cells where YK:075c is either not tagged or tagged with the Myc epitope by immunofluorescence using anti-Myc antibodies and AlexaFluor488-tagged secondary antibodies. Scale bar represents 2μm. (b) A volcano plot displaying the transcriptional changes between WT and *ykl075cΔ* cells from 2 RNA sequences. Y-axis plots significant and x-axis is fold change. (c) Table lists BCAA proteins with altered transcript levels in *ykl075cΔ* cells. (d-e) Validating mRNA transcript levels of BCAA transaminases, *BAT1* and *BAT2* through quantitative PCR. Decreased *BAT1* and increased *BAT2* gene expression in *ykl075cΔ* cells are significantly different from WT cells using unpaired two-tailed T-test, **p<0.01. Data presented is a representative trial from 3 experiments. (f) Quantification of free intracellular BCAA levels in WT, *ykl075cΔ*, *bat1Δ*, and *ykl075cΔbat1Δ* cells using Cell BioLabs BCAA colorimetric elisa kit. Intracellular free BCAA levels were calculated based on a leucine standard and normalized to protein concentration isolated from each strain. Data values are represented from 3 trials and statistically analyzed using One-way Anova test (*p<0.05, **p<0.01, ***p<0.001 ****p <0.0001). (g) Representative trial images from YPD plates of a serial growth dot-assay of WT, *ykl075cΔ*, *tor1Δ*, and *ykl075cΔtor1Δ* cells recovering from either DMSO or 200nM rapamycin treatment. (h) Optical density per hour measurements of WT, *ykl075cΔ*, *tor1Δ*, and *ykl075cΔtor1Δ* growth rate during recovery from DMSO or 200nM rapamycin treatment.

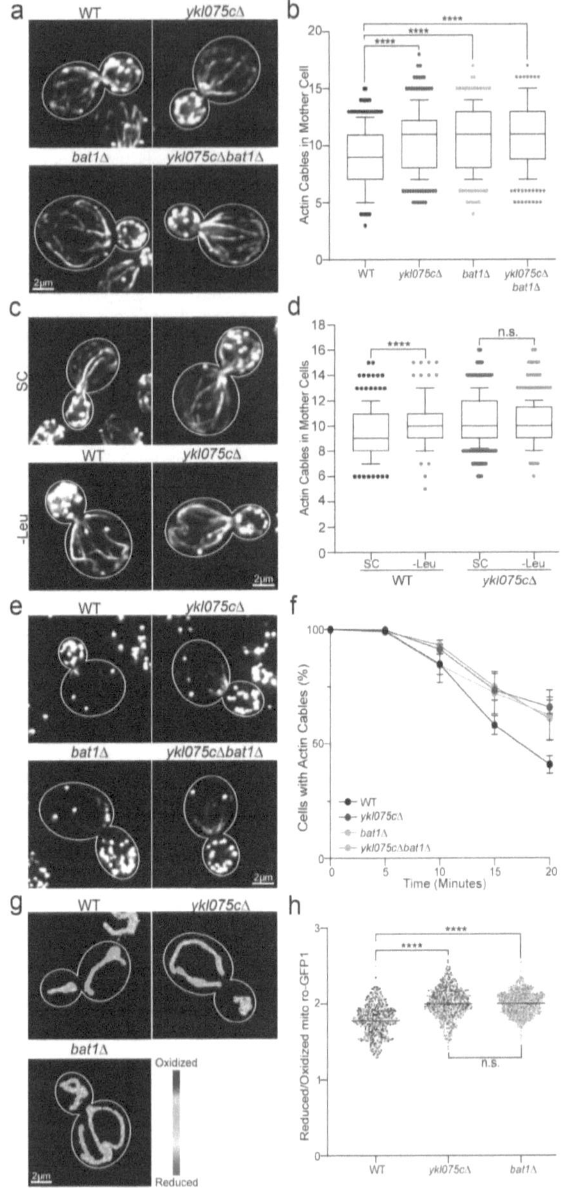

a
WT
ykl075cΔ
bat1Δ
ykl075cΔbat1Δ
2μm
b
20
15
10
5
0
Actin Cables in Mother Cell
****
****
****
WT
ykl075cΔ
bat1Δ
ykl075cΔ
bat1Δ
c
SC
WT
ykl075cΔ
-Leu
2μm
d
18
16
14
12
10
8
6
4
2
0
Actin Cables in Mother Cells
****
n.s.
SC
-Leu
WT
SC
-Leu
ykl075cΔ
e
WT
ykl075cΔ
bat1Δ
ykl075cΔbat1Δ
2μm
f
100
50
Cells with Actin Cables (%)
WT
ykl075cΔ
bat1Δ
ykl075cΔbat1Δ
0
5
10
15
20
Time (Minutes)
g
WT
ykl075cΔ
bat1Δ
Oxidized
Reduced
2μm
h
3
2
1
0
Reduced/Oxidized mito ro-GFP1
****
****
n.s.
WT
ykl075cΔ
bat1Δ

Figure 2.4: *YKL075C* **affects actin cables and mitochondrial through its function in BCAA homeostasis.**

(**a**) Representative images of AlexaFluor488 phalloidin-stained actin cytoskeleton of mid-log phase WT, *ykl075cΔ*, *bat1Δ*, *ykl075cΔbat1Δ* cells. (**b**) Box & whiskers plots showing actin cable abundance in the respective strains (n>230 cells/strain from 3 trials). The central band represents the median, and the boxes indicated the middle quartiles; whiskers extend to the 10th and 90th percentiles, and points indicate outliers. ****p <0.0001 (non-parametric Kruskal-Wallis test). (**c**) Representative images of AlexaFluor488 phalloidin-stained actin cytoskeleton of mid-log phase WT, *ykl075cΔ*, *bat1Δ* or *ykl075cΔbat1Δ* cells grown in synthetic complete (SC) medium or SC without leucine (-Leu). Scale bar: 2μm. (**d**) Actin cable abundance shown in a box & whiskers graph. The central band represents the median, and the boxes indicated the middle quartiles; whiskers extend to the 10th and 90th percentiles (n>70 cells/strain/trials). Data represents 3 trials. ****p <0.0001 (non-parametric Kruskal-Wallis test). (**e**) Representative images of AlexaFluor488 phalloidin-stained actin cytoskeleton of mid-log phase WT, *ykl075cΔ*, *bat1Δ*, *ykl075cΔbat1Δ* cells treated with 2μM Lat-A for 20 min. Scale bar: 2μm. (**f**) Actin cable loss as a function of the time of Lat-A treatment in WT, *ykl075cΔ*, *bat1Δ*, *ykl075cΔbat1Δ* cells. Error bars: SEM of 3 trials (n = 100 cells/strain/trial). WT vs. *ykl075cΔ*: p-value = 0.0013, WT vs. *bat1Δ*: p-value = 0.0363, and WT vs. *yk075cΔbat1Δ*: p-value = 0.0199. p-values were determined by simple linear regression analysis. (**g**) Representative images of mitochondria redox state detected by using mito-roGFP1. Scale bar: 2μm. (**h**) Reduced:oxidized mito-roGFP1 ratios in WT, *ykl075cΔ*, *bat1Δ*, and *ykl075cΔbat1Δ* cells (n>250 cell/strain from 3 trials) were determined as for Fig. 2.1g. The central band represents the median. ****p <0.0001 (non-parametric Kruskal-Wallis test)

6   Materials and Methods

Yeast growth conditions

All *S. cerevisiae* strains used in this study are derivatives of the wild-type BY4741 strain (*MATa his3Δ1 leu2Δ0 met15Δ0 ura3Δ0*) or S288C strain (*MATα SUC2 gal2 mal2 mel flo1 hap1 ho bio1 bio6*) from Open Biosystems (Hunstville, AL). All cultures used in imaging and RLS methods were grown to mid-log phase (OD_{600} 0.1-0.5) by shaking at 30°C or 25°C in either rich glucose-based medium, yeast-peptone-dextrose (YPD) or a nitrogen based medium, synthetic complete (SC) medium.

Yeast strain construction

Knock strains (Supplemental Table 2.1) were created using homologous recombination. Genes of interests were replaced with either auxotrophic marker or drug marker cassettes using appropriate primers. Auxotrophic marker cassettes used were *URA3* and *LEU2*. Drug cassettes used were *kanMX6* or *hphMX6* from pFA6-KanMX6 and PCY3090-02, respectively. For example, CSY036 was created by replacing genomic *YKL075C* with KanMX6. A PCR fragment containing regions homologous to sequences directly upstream and downstream of the start and stop codons of *YKL075C* and coding regions for KanMX6 was amplified from plasmid pFA6-KanMX6 (Addgene, Cambridge, MA) using primers listed in Supplemental Table 2.2. Cells then were transformed with PCR product using the lithium acetate method and transformants were selected on appropriate selective media plates with 200µg ml^{-1} Geneticin (Sigma-Aldrich, St Louis, MO) to select for KanMX6-positive cells. Positive cells were confirmed by PCR sequencing.

The same techniques as above were used to add fluorescent protein tags to proteins at their chromosomal loci using PCR fragments amplified from plasmids: pFA6a-GFP(S65ST)-KanMX6 and pFA6a-GFPEnvy-KanMX6 (Addgene). Selection for positive transformants was carried out as described above and visually confirmed by wide-field fluorescence microscopy.

ACT1-V159N mutation was generated in BY4741 using CRISPR [231, 241]. Briefly, pML104 containing both Cas9 the guide RNA expression cassette is linearized by digestion with BclI and SwaI enzymes. Oligonucleotides containing a GATC overhang, a 20bp guide sequence immediately preceding a PAM sequence near the point mutation site, and the 5' end of the structural segment of the single guide RNA (sgRNA) were designed and hybridized. The hybridized oligonucleotides were ligated into the digested pML104 plasmid to generate the complete sgRNA expression cassette. Repair oligonucleotides containing the *V159N* point mutation, PAM sequence mutation that eliminates the PAM sequence but does not change amino acid sequence, and guidance sequence flanked by 40bp of homology to regions immediately upstream and downstream of the guidance sequence were generated and hybridized. Yeast cells were transformed with both pML104 containing complete sgRNA cassette and repair oligonucleotides, and the transformants were selected on SC-Ura plates. Positive transformants were confirmed by sequencing.

Microscopy

Fluorescence microscopy were acquired by wide-field imaging and 3D super-resolution structured illumination imaging. Images are acquired on the respective microscopes. Axioskop 2 microscope equipped with 100x/1.4 Plan-Apochromat objective (Zeiss) and an Ocra-ER cooled CCD camera (Hamamatsu).). Fluorescence was excited by a pE-4000 LED illumination system (cooLED, Andover, UK). The system was controlled by NIS Elements 4.60 Lambda software (Nikon, Mellville, NY). Nikon super-resolution structured illumination microscope was equipped with 100x/1.45 PSF oil immersion objective lens (Nikon) and EM-CCD Camera controlled by NIS Elements 4.60 Lambda software (Nikon, Mellville, NY). Details are given with each imaging methods.

Visualization of F-actin Cytoskeleton and bud scars

Cells at mid-log phase were concentrated to an OD_{600} of 0.5 and resuspended in 1ml of growth medium. Cells were fixed in 3.7% paraformaldehyde added directly to 1ml of cells at 30°C with shaking for 50 min. Fixed cells were washed three times with 1xPBS, followed by one wash with 1xPBT (PBS containing 1% w/v BSA, 0.1% v/v Triton X-100, 0.1% w/v sodium azide), and actin was stained with 2.5µM AlexaFluor 488-phalloidin (Thermo Scientific, Grand Island, NY) for 25 min at room temperature in the dark. Cells were then washed three times with 1xPBS, and resuspended in Sigma prolong anti-fade mounting solution.

Visualization of stained actin cytoskeleton was carried out by wide-field fluorescence microscopy. Stained cells were mounted in 1.8µl SlowFade diamond antifade mountant (Thermo Scientific, Grand Island, NY) on a glass slide with a #1.5 coverslip. AlexaFluor 488 phalloidin was excited with a 470nm LED and an ET470/40x filter (Chroma, Bellows Falls, VT). Z stacks were captured at 0.3µm intervals using 1x1 binning, 200-300ms exposure time, and gain of 116. Images were deconvolved using a constrained iterative restoration algorithm assuming 507nm emission wavelength with a 100% confidence limit and 60 iterations, using Volocity 5.5 (Quorum Technologies, Guelph, Ontario, Canada). All images were contrast-enhanced with similar parameters.

Cells visualized with SIM were fixed and stained as previously described. Additional staining for bud scars required 5µg/ml WGA594 (Thermo Scientific, Grand Island, NY). Cells were prepared following protocol described in Higuchi-Sanabria et al., 2016 [242]. Prepared slides were cured overnight in Prolong diamond antifade mountant (Thermo Scientific, Grand Island, NY) in dark at room temperature before imaging. SIM images were captured on Nikon Blank scope with 100x/1.45 PSF oil immersion objective lens (Nikon) and EM-CCD Camera with 200 gain. Alexa488 phalloidin was excited with 488nm laser at 45% power with 300ms exposure time. WGA594 was excited with 561nm laser at 10% power with 200ms exposure time. Z-series were collected through entire cell at 0.125µm intervals using 1x1 binning. SIM images were then reconstructed using parameters described in Higuchi-Sanabria et al. [242]. All image were contrast-enhanced with similar parameters in each channel.

Immunofluorescence

Fix 5ml culture of mid-log growth phase yeast cells (0.2-0.5 OD_{600}) with 3.7% final paraformaldehyde concentration for 1 h shaking in 30°C incubator. Executed protocol for immunofluorescence in Higuchi-Sanabria 2016 et al actin imaging methods chapter. Lastly, used 1.5µl of SlowFade mountant (Thermo Scienific, Grand Island, NY) to adhere cells to the microscope slide. Seal coverslip edges with slide sealer and image or store in 4°C.

Analysis of actin cable stability

Treat cells with 2µM of Latrunculin-A (Sigma-Aldrich, St. Louis, MO). Cells were either incubated with Lat-A for 3-days during growth curve analysis or treated for 20 min to determine existence of actin cables using Alexa488 phalloidin staining. Determining growth rates required manual readings OD_{600} for 3-days at 1 h intervals. Analysis of actin cables post-Lat-A treatment was conducted on fixed cells, as previously described, at 5 min intervals. Acquired images analyzed using the same parameters was described above.

Analysis of Mitochondrial Redox State using mito-roGFP1

Imaging mitochondrial redox state using mito-roGFP1 was performed as previously described in McFaline-Figueroa et al. and Vevea et al. [71, 233]. However, cells were either transformed with plasmid bearing mito-roGFP1 sequence or integrated mito-roGFP1 at Ho locus. Images were acquired on wide-field microscope previously described above. Reduced and oxidized states were excited at 470nm and 365nm with a GFP filter (Zeiss filter 46HE, dichroic FT 515, emission 535nm/30) without an excitation filter. 0.3µm intervals in the Z-axis were captured among the entire cell depth. Captured images were deconvolved using constrained 60 iterative restoration algorithm with a calculated PSF using the following parameters: λ_{ex} 507nm and 100% confidence limit (Volocity 5.5, Quorum Technologies). After subtracting background and thresholding, the reduced/oxidized ratio was calculated by dividing the intensity of the reduced channel (λ_{ex} = 470nm, λ_{em} = 525nm) by the intensity of the oxidized channel (λ_{ex} = 365nm, λ_{em} = 525nm) in Volocity software.

RLS Analysis

RLS measurements were performed as previously described in [153], without alpha-factor synchronization. Selected strains stored at -80°C were streaked out on YPD plates and grown for 2-days at 30°C. A patch of colonies were selected and grown overnight in liquid YPD at 30°C to mid-log, exponential, growth phase. 10µl of cell suspension was streaked onto a YPD plate. Small-budded cells were isolated and arranged in a matrix using a micromanipulator mounted on a dissecting microscope (Sporeplay, Singer Instruments, Somerset, UK). Upon a complete budding event, mother cells were removed and discarded to leave only virgin cells. The time and number of subsequent daughter cells produced by each virgin mother cell were recorded until all replication ceased.

Determining branched-chain amino acid levels

Extract cell lysate from 8-10 OD_{600} of mid-log growth phase cells by vortex in cells with 0.5mm glass beads in chilled 1xPBS for 5 min. Spin down and transfer supernatant to ice. Measure total protein concentration using standard Pierce™ BCA protein assay kit (Thermo Scientific, Grand Island, NY). Then followed Branched Chain Amino Acid Assay kit manual given by Cell Biolabs, Inc (San Diego, CA). Measured BCAA levels in triple technical replicates against negative control samples lacking leucine dehydrogenase. Levels of free BCAAs extracted are normalized to extracted protein concentration from cell lysates.

RNA Sequencing

Transcriptome was analyzed as previously described in Liao et al. [243]. RNA was extracted from mid–log phased WT and *ykl075cΔ* cells using the RNeasy kit (Qiagen, Germantown, MD). RNA quality was analyzed with Agilent 2100 Bioanalyzer using a Plant RNA Nano chip and RNA Integrity Number (RIN) scores higher than 9 were sequenced. RNA library preparations and sequencing reactions were conducted at GENEWIZ, LLC. (South Plainfield, NJ, USA). RNA sequencing libraries were prepared using the NEB Next Ultra RNA Library Prep Kit for Illumina using manufacturer's instructions (NEB, Ipswich, MA, USA). The sequencing library was validated on the Agilent Tape Station (Agilent Technologies, Palo Alto, CA, USA), and quantified using a Qubit 2.0 Fluorometer (Invitrogen, Carlsbad, CA) as well as by quantitative PCR (KAPA Biosystems, Wilmington, MA, USA). The sequencing libraries were clustered on a single lane of a flowcell. After clustering, the flowcell was loaded on the Illumina HiSeq instrument 4000 according to manufacturer's instructions. The samples were sequenced using a 2x150bp Paired End (PE) configuration. Sequence reads were trimmed to remove possible adaptor sequences and nucleotides with poor quality using Trimmomatic v.0.36. The trimmed reads were mapped to the Saccharomyces cerevisiae S288C reference genome available on ENSEMBL using the STAR aligner v.2.5.2b. Unique gene hit counts were calculated by using feature counts from the Subread package v.1.5.2. After extraction of gene hit counts, the gene hit counts table was used for downstream differential expression analysis. Using DESeq2, a comparison of gene expression between WT and *ykl075cΔ* cells was performed. The Wald test was used to generate p-values and $\log_2$ fold changes. Genes with a p-value < 0.05 and absolute $\log_2$ fold change > 1 were called as differentially expressed genes. REVIGO was used to remove redundant GO terms and group-related GO terms in semantic similarity-based scatterplots [244].

cDNA synthesis and quantitative PCR

Extracted RNA from cultured mid-log phase WT or *ykl075cΔ* cells using the RNeasy kit (QIAGEN, Germantown, MD). Genomic DNA contamination was removed using TURBO DNA-free Kit (Ambion, Carslbad, CA). 1 mg of DNA-free RNA was used for cDNA synthesis with SuperScript IV First-Strand Synthesis System (Invitrogen, Waltham, MA). cDNA was diluted and used for quantitative PCR reaction with PowerUp SYBR Green Master Mix (Applied Biosystems, Carlsbad, CA). Primers for qPCR were designed using NCBI Primer Blast (https://www.ncbi.nlm.nih.gov/tools/primer-blast/) with a PCR product size of 100 bp and max Tm difference of 2C. For each specified gene, ΔCT was calculated as CT_{gene} - CT_{actin}, and fold change was calculated as $2^{-\Delta\Delta CT}$ with actin serving as the endogenous control for each sample.

ACT1 was amplified with 5′ TCGCCTTGGACTTCGAACAA 3′ and 5′ CAAAGCTTCTGGGGCTCTGA 3′,

p195 BAT1_RT-qPCR_F   5'ACCCAAATCCATCCAAGCCA 3'

p196 BAT1_RT-qPCR_R   5' CACCCTTCTTTGGCTGACCA 3'

p197 BAT2_RT-qPCR_F   TTTCCACGCCTGATAGAGCC

p198 BAT2_RT-qPCR_R   CAGTGGCTTCCAGTCTGACC

Quantification and Statistical Analysis

All data were analyzed for normal distribution with the D'Agostino and Pearson normality test. p-values for simple two-group comparison were determined with a two-tailed Student's t-test for parametric distributions and a Mann-Whitney test for non-parametric data. For multiple group comparisons, p-values were determined by a 1-way ANOVA with Dunnett's or Sidak's test for parametric distributions and a Kruskal-Wallis test with Dunn's post hoc test for non-parametric distributions. Survival curves were statistically compared using a Log-rank (Mantel-Cox) test. GraphPad Prism7/8 (GraphPad Software) was used to conduct all statistical analysis. Bar graphs show the mean and SEM; in box and whisker graphs, the box represents the middle quartile, the midline represents the median and whiskers show the 10th and 90th. For all tests, p-values are classified as follows: *p < 0.05; **p < 0.01; *** p < 0.001; ****p < 0.0001.

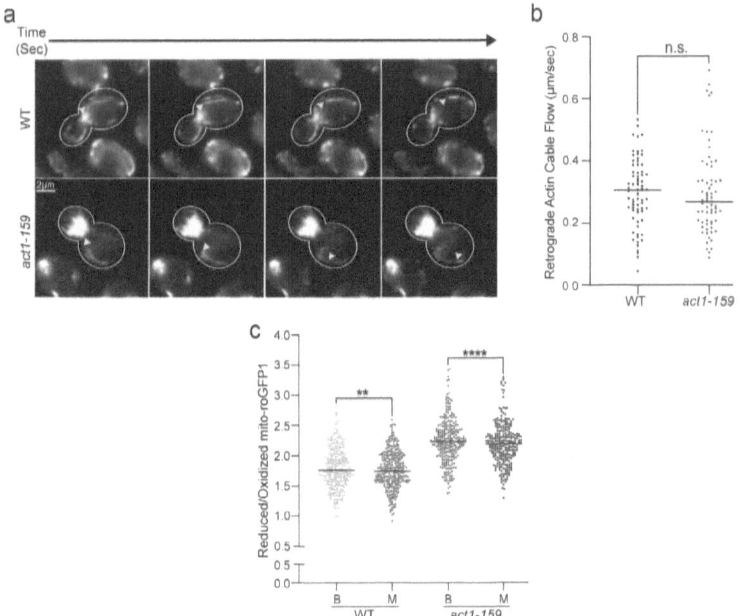

Supplementary Figure 2.1: Increasing actin cable stability does not affect RACF or asymmetric inheritance of mitochondria.
(a) Still frames from a time-lapse series of the retrograde flow actin cables visualized using Abp140p labeled at its chromosomal locus with GFPEnvy. Arrowheads indicate motile actin cable. (b) The velocity of RACF in WT and *act1-159* cells (n>60 cells). n.s. – not significant (Mann-Whitney test). (c) Quantifications of reduced:oxidized mito-roGFP1 ratios in buds and mother cells of WT and *act1-159* cells (n> 250 cells/strain from 3 trials). ****p<0.0001, using Wilcoxon matched-pair test.

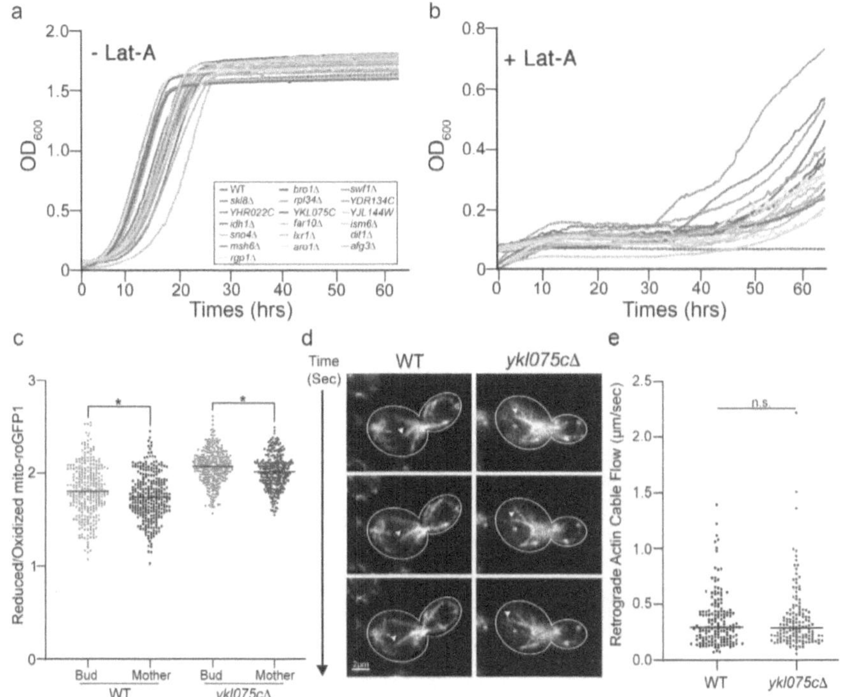

Supplementary Figure 2.2: Analysis of RACF and mitochondrial redox state in *ykl075cΔ* cells.
(**a-b**) Growth rate curves measuring by OD_{600} treated with and without Lat-A. Graph plots top hits from genome-wide screen treated with Lat-A whose deletion had increased resistance to growth defects from Lat-A. (**c**) Reduced:oxidized mito-roGFP1 ratios in mitochondria in mother cells and buds of WT and *ykl075cΔ* cells (n>250 cells/strain from 3 trials). ****$p<0.0001$ (Wilcoxon matched-pair test). (**d**) Still frames from a time-lapse series of actin cable marker Abp140p-GFP undergoing retrograde actin cable flow (RACF) in mid-log phase WT and *ykl075cΔ* cells. Arrowheads points to a motile actin cable. (**e**) RACF rates in WT and *ykl075cΔ* cells (n>140 cells/strain for 3 trials). n.s. – not significant (non-parametric Mann-Whitney test).

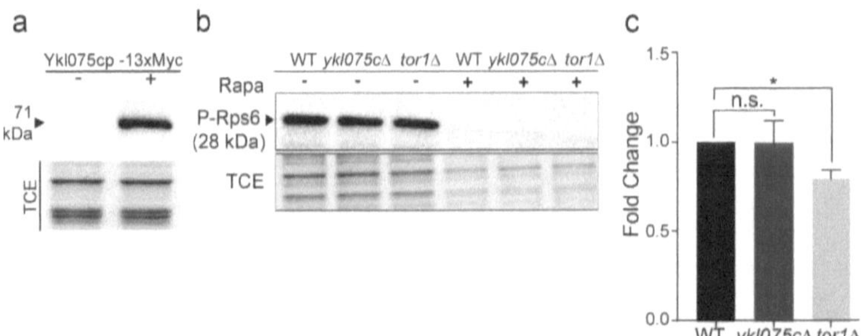

Supplementary Figure 2.3: *YKL075C* functional role in the actin cytoskeleton is independent of the TORC1 pathway

(a) Western blot of C-terminally 13-Myc tagged Ykl075cp. TCE, total protein loading control. (b) Representative western blot staining for phospho-levels of ribosomal protein-S6 in WT, *ykl075cΔ* and *tor1Δ* cells with and without rapamycin (200μM). Lower panel used TCE staining as a loading control. (c) Quantification phospho-levels from immunoblot of mid-log phase WT, *ykl075cΔ* and *tor1Δ* cells from 5 trials. *p<0.05 (non-parametric Kruskal-Wallis test).

64

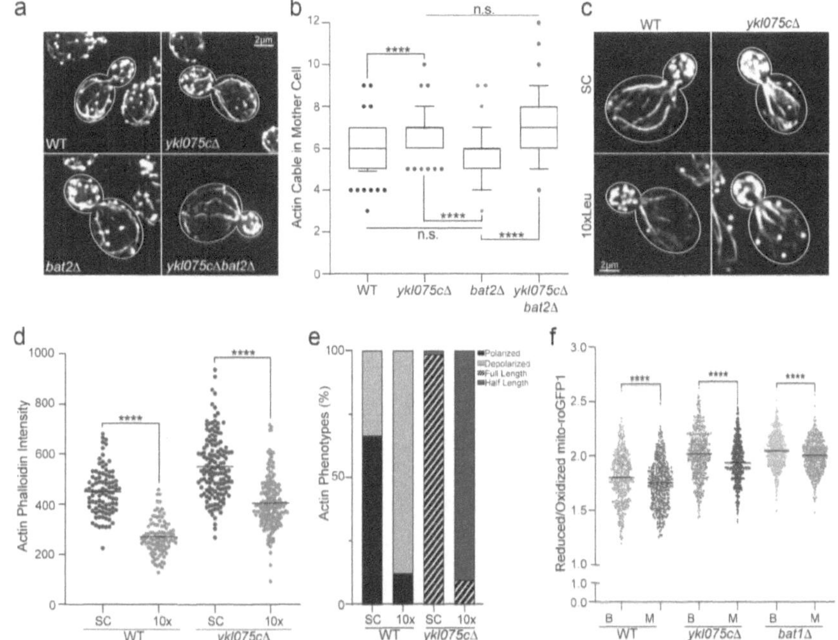

Supplementary Figure 2.4: Deletion of *BAT2* elicits no change in actin cable abundance, while leucine supplement affects actin cable abundance and polarity.

(a) Representative images of AlexaFluor488 phalloidin-stained actin cytoskeleton of mid-log phase WT, *ykl075cΔ*, *bat2Δ*, or *ykl075cΔbat2Δ* cells. (b) Actin cable abundance shown in a box & whiskers graph of WT, *ykl075cΔ*, *bat2Δ*, or *ykl075cΔbat2Δ* cells. The central band represents the median, and the boxes indicated the middle quartiles; whiskers extend to the 10th and 90th percentiles (n>70 cells/strain/trials). Data are representative of 3 trials. ****p <0.0001 (non-parametric Kruskal-Wallis test). (c) Representative images of AlexaFluor488 phalloidin-stained actin cytoskeleton of mid-log phase WT and *ykl075cΔ* cells grown in synthetic complete (SC) medium or SC with leucine supplementation. Scale bar: 2μm. (d) Quantification F-actin phalloidin signal intensity in mother cells of WT and *ykl075cΔ* strains. The central band represents the median. Data are representative from 3 trials (n>95 cells per strain/trials). ****p<0.0001 (non-parametric Kruskal-Wallis test). (e) Actin cytoskeleton phenotypes in mid-log phase of WT and *ykl075cΔ* grown in SC or SC with 10X leucine. (f) Reduced:oxidized mito-roGFP1 ratios of buds and mother cells of mid-log phase WT, *ykl075cΔ*, and *bat1Δ* cells (n>250 cells/strain from 3 trials). ****p<0.0001 (Wilcoxon matched-pair test).

Strains	Genotype	Source
CSY064	MATa his3Δ1 leu2Δ0 met15Δ0 ura3Δ0	This Study
CSY032	MATα SUC2 gal2 mal2 mel flo1 hap1 ho bio1 bio6	This Study
CSY036	MATα SUC2 gal2 mal2 mel flo1 hap1 ho bio1 bio6 ykl075cΔ::kanMX-6	This Study
CSY046	MATa his3Δ1 leu2Δ0 met15Δ0 ura3Δ0 TOM70-mCherry hphMX-4 [pmito-roGFP1:URA3]	This Study
CSY065	MATα SUC2 gal2 mal2 mel flo1 hap1 ho bio1 bio6 tor1Δ::hphMX-4	This Study
CSY074	MATα SUC2 gal2 mal2 mel flo1 hap1 ho bio1 bio6 ykl075cΔ::kanMX-6 tor1Δ::hphMX-4	This Study
CSY093	MATα SUC2 gal2 mal2 mel flo1 hap1 ho bio1 bio6 ykl075cΔ::hphMX-4 [pmitoHo-roGFP1:KanMX]	This Study
CSY104	MATα SUC2 gal2 mal2 mel flo1 hap1 ho bio1 bio6 [pmitoHo-roGFP1:KanMX]	This Study
CSY107	MATα SUC2 gal2 mal2 mel flo1 hap1 ho bio1 bio6 ykl075cΔ::hphMX-4 [pmitoHo-roGFP1:KanMX]	This Study
CSY113	MATα SUC2 gal2 mal2 mel flo1 hap1 ho bio1 bio6 ykl075cΔ::hphMX-4 ABP140-GFP (S65T) kanMX-6	This Study
CSY125	MATα SUC2 gal2 mal2 mel flo1 hap1 ho bio1 bio6 ABP140-GFP (S65T) kanMX-6	This Study
CSY152	MATa his3Δ1 leu2Δ0 met15Δ0 ura3Δ0 ABP140-GFP (S65T) kanMX-6	This Study
CSY154	MATa his3Δ1 leu2Δ0 met15Δ0 ura3Δ0 act1-V159N [pML104]	This Study
CSY160	MATα SUC2 gal2 mal2 mel flo1 hap1 ho bio1 bio6 bat1Δ::hphMX-4	This Study
CSY164	MATα SUC2 gal2 mal2 mel flo1 hap1 ho bio1 bio6 ykl075cΔ::kanMX-6 bat1Δ::hphMX-4	This Study
CSY183	MATa his3Δ1 leu2Δ0 met15Δ0 ura3Δ0 act1-V159N [pML104] [pmito-roGFP1:URA3]	This Study
CSY186	MATa his3Δ1 leu2Δ0 met15Δ0 ura3Δ0 ykl075cΔ::LEU2	This Study
CSY187	MATa his3Δ1 leu2Δ0 met15Δ0 ura3Δ0 act1-V159N ABP140-ENVY his3 [pML104]	This Study
CSY225	MATα SUC2 gal2 mal2 mel flo1 hap1 ho bio1 bio6 YKL075C-13Myc-kanMX-6	This Study
EGS164	MATa his3Δ1 leu2Δ0 met15Δ0 ura3Δ0 ABP140-ENVY his3	This Study
RHY055	MATa his3Δ1 leu2Δ0 met15Δ0 ura3Δ0 ykl075cΔ::LEU2	This Study
CTY056	MATα SUC2 gal2 mal2 mel flo1 hap1 ho bio1 bio6 bat1Δ::hphMX-4 [pmitoHo-roGFP1:KanMX]	This Study

Supplementary Table 2.2:

Strains	Genotype	Source
FW YKL deletion pFA6	TAACATTATAAATCAAGCCTAGTATAGCGCGCCAAAGAA Gcggatccccgggttaattaa	This Study
RV YKL deletion pFA6	AGAAAACGCTGGATCTTGCTTCTTTTCCTTTGATTTCTCC gaattcgagctcgtttaaac	This Study
FW YKL deletion PCY	TAACATTATAAATCAAGCCTAGTATAGCGCGCCAAAGAA Ggtttagcttgcctcgtcc	This Study
RV YKL deletion PCY	AGAAAACGCTGGATCTTGCTTCTTTTCCTTTGATTTCTCC atcgatgaattcgagctcg	This Study
FW ABP140 Tag pFA6 GFP	GTACCGCTGCTGGGTACAAGCTGTGTTTGACGTTCCTCA Acggatccccgggttaattaa	This Study
RV ABP140 Tag pFA6 GFP	TTTATGATGAGAGAGGAGGTGGTACTTGTCTCAGAACTT Cgaattcgagctcgtttaaac	This Study
sgFW ACT1-159	GATCTAGAACTACTGGTATTGTTTGTTTTAGAGCTAG	This Study
sgRV ACT1-159	CTAGCTCTAAAACAAACAATACCAGTAGTTCTA	This Study
FW RT ACT1-159	TTGTCCTTGTACTCTTCCGGTAGAACTACTGGTATTGTTT TAGATTCCGGTGATGGTAATACTCACGTCGTTCCAATTT ACGCTGGTTTCTCTCTACCTCAC	This Study
RV RT ACT1-159	GTGAGGTAGAGAGAAACCAGCGTAAATTGGAACGACGT GAGTATTACCATCACCGGAATCTAAAACAATACCAGTAG TTCTACCGGAAGAGTACAAGGACAA	This Study
FW BAT1 deletion PCY	TATAAACGCAAAATCAGCTAGAACCTTAGCATACTAAAAC gtttagcttgcctcgtcc	This Study
RV BAT1 deletion PCY	TTTTTTTTGGGGGGGGGAGGGGATGTTTACCTTCATTATC Atcgatgaattcgagctcg	This Study
FW YKL deletion pOM	TAACATTATAAATCAAGCCTAGTATAGCGCGCCAAAGAA Gtgcaggtcgacaacccttaat	This Study
RV YKL deletion pOM	AGAAAACGCTGGATCTTGCTTCTTTTCCTTTGATTTCTCC gcagcgtacggatatcaccta	This Study
FW ABP140 Tag pFA6 GFPEnvy-linker	GTACCGCTGCTGGGTACAAGCTGTGTTTGACGTTCCTCA Aggtgacggtgctggtttaattaac	This Study
RV ABP140 Tag pFA6 GFP	TTTATGATGAGAGAGGAGGTGGTACTTGTCTCAGAACTT Cgaattcgagctcgtttaaac	This Study
FW YKL pFA6 13Myc Tag	GGAAAAAATATGTGAAGTTGCTGCAAGACAGAAATGGAT Acggatccccgggttaattaa	This Study
RV YKL pFA6 13Myc Tag	AGAAAACGCTGGATCTTGCTTCTTTTCCTTTGATTTCTCC gattcgagctcgtttaaac	This Study

Supplemental Table 2.3:

Upregulated Gene List of 2 RNAseqs (A&B)			
Alias	Log2Fold Change	p-Value	RNA-Seq Trial
ECM23	3.509835254	0.01838047	A
YRF1-1	2.679760255	0.03135367	A
HED1	2.012335594	0.007184716	A
YBR300C	1.911933697	0.021288226	A
YAT1	1.746750599	1.55508E-12	A
FRE7	1.538823129	7.73564E-15	A
COS2	1.317190176	0.035306513	A
YGR139W	1.283499963	3.09355E-09	A
YHR054C	1.244823199	0.003381491	A
tD(GUC)N	1.238533232	0.019786784	A
PRR2	1.161684234	3.99117E-05	A
YNL054W-B	1.152867881	6.43784E-08	A
UBC11	0.991241563	0.021313901	A
HXT9	0.983766442	0.020402829	A
YHL050C	0.976961264	0.000167045	A
YJR149W	0.943789966	0.00076322	A
YOR338W	0.916835434	0.000467125	A
FUS2	0.828875069	0.011506276	A
REC8	0.772230025	0.021666125	A
YIR018C-A	0.761374021	0.045306979	A
SPS18	0.753760837	0.02486144	A
YGL081W	0.740997885	0.035151512	A
RPL15a	0.706958844	0.003980299	A
YBL113C	0.700686878	0.002547115	A
UGA1	0.696495551	0.00025099	A
YLR410W-B	0.681622999	0.001032075	A
COS8	0.673048238	0.000419257	A
POR1	0.66427765	0.002466703	A
REE1	0.662181642	0.000493582	A
IRC7	0.656712523	0.000559798	A
YDR541C	0.651926047	0.005782276	A
NAR1	0.65179809	0.001255985	A
YAP5	0.65082168	0.004868877	A
PDC5	0.649855051	0.004937099	A
FLO10	0.64228555	0.005521326	A
ATG32	0.641123407	0.004832691	A
IMA2	0.634912387	0.038817465	A
ADR1	0.624297472	0.001314558	A
CHA4	0.617281544	0.001264685	A
SIT1	0.598731134	0.002180914	A
CYC1	0.596640081	0.00773712	A
CHA1	0.575199405	0.007967885	A
Ho	0.570653323	0.007807777	A
FRE4	0.550584386	0.006000584	A
MIG2	0.533966916	0.008317386	A
STP4	0.529924047	0.010120158	A
CTR1	0.529154476	0.011354663	A
YHR033W	0.525984994	0.011746041	A
IMD2	0.525842308	0.009772665	A
YER053C-A	0.52537212	0.008989395	A

YGR079W	0.524601965	0.043180279	A
IDH2	0.519491517	0.007963545	A
ODC1	0.514139668	0.045571256	A
TRA1	0.500558919	0.026961272	A
FIG1	1.90718932	0.024766877	B
tD(GUC)N	1.363221566	0.037893273	B
CRC1	1.307598045	0.04798266	B
INH1	1.184460297	0.017973344	B
YGL081W	1.135660891	0.0310327	B
PHM7	1.065101634	0.028991975	B
COX26	1.05777735	0.044931963	B
CYC7	1.029834879	0.024015136	B
GIP2	1.013992581	0.026497207	B
CHA4	0.989410647	0.023228096	B
YCR062W	0.989148038	0.022515387	B
YAT1	0.986516483	0.02657154	B
TMA10	0.968830784	0.022475358	B
URC2	0.961183689	0.024933329	B
RTN2	0.928456327	0.027485132	B
GPH1	0.909605359	0.02531816	B
CYC1	0.903234482	0.030250572	B
STP4	0.886078595	0.025947736	B
CYT1	0.882088021	0.028316009	B
PLB3	0.880779365	0.026804349	B
FMP48	0.878571708	0.028445848	B
RTC3	0.867829138	0.027749353	B
FRE7	0.857294229	0.029986167	B
YCL042W	0.847204172	0.034296472	B
YRO2	0.83800556	0.028881586	B
UGA1	0.809162078	0.032305836	B
OM14	0.802837145	0.033929902	B
YNL194C	0.792623103	0.040568814	B
SDH1	0.788717247	0.035164191	B
IMD2	0.779023023	0.034414629	B
ASG7	0.770986038	0.045112631	B
SPI1	0.757191104	0.035707006	B
USV1	0.757145726	0.038804829	B
TPO2	0.747992559	0.036203533	B
YNL190W	0.732574419	0.037910796	B
SSA4	0.730736723	0.038695007	B
PAN1	0.725855711	0.0384879	B
YNR014W	0.717062666	0.041605156	B
POR1	0.715918058	0.038941782	B
SRO9	0.700293758	0.046027774	B
RPL15A	0.682808339	0.042324276	B
TSL1	0.679495673	0.042969537	B
SED1	0.67929313	0.042781594	B
MSC1	0.658682729	0.047510232	B
PNS1	0.652196787	0.049102765	B
GAD1	0.647803909	0.04818775	B

Supplementary Table 2.4:

Downregulated Gene List of 2 RNAseqs (A&B)			
Alias	**Log2Fold Change**	**p-Value**	**RNA-Seq Trial**
ARO10	-0.739325719	0.000186362	A
ARO9	-0.826750506	0.000494406	A
BAP2	-0.894707252	8.24521E-05	A
BAT1	-1.506299101	1.5104E-12	A
ILV2	-0.74555263	0.000575787	A
ILV3	-0.659641914	0.002074562	A
ILV5	-0.615666144	0.005746103	A
ISU2	-1.102793146	7.47341E-06	A
LEU1	-2.450496479	1.19143E-27	A
LEU4	-0.557400301	0.003937671	A
PHO8	-0.576946531	0.008095742	A
SNR8	-1.254045342	2.79691E-07	A
YBL029C-A	-0.951560934	3.32669E-06	A
YKL075C	-9.190604464	1.2019E-104	A
AGP1	-0.581577719	0.002213802	A
AIM29	-0.751767378	0.000298058	A
ALD5	-1.344159103	4.46592E-10	A
ATX1	-0.912059587	1.23601E-05	A
BLS1	-0.66411176	0.026347458	A
BUD26	-0.833965022	0.000347444	A
CHO2	-0.507009732	0.020879817	A
CIT2	-0.664745329	0.000483054	A
CNL1	-0.673687351	0.001243064	A
COX17	-0.501986951	0.012487515	A
COX7	-0.573024238	0.002732459	A
CSN9	-0.543656153	0.017870691	A
DAD2	-0.568733799	0.00752023	A
DAD3	-0.534019036	0.010773402	A
DAD4	-0.877815908	6.00724E-06	A
DIC1	-0.805070786	0.000543911	A
ECL1	-0.726902408	0.000128706	A
EFG1	-0.604351844	0.001462238	A
EMC6	-0.696385081	0.000260453	A
ENA1	-0.752153342	0.000528544	A
FUI1	-0.511830198	0.007421294	A
GAT2	-0.607567289	0.002203451	A
GDH1	-0.854773886	4.50191E-05	A
GID8	-0.559076451	0.003350961	A
GIM4	-0.846348446	0.000129639	A
HMRA1	-0.718918803	0.000596365	A
HSP78	-0.73948367	0.000849159	A
HTL1	-0.592133296	0.007804204	A
ICY2	-0.774742754	4.71203E-05	A
IST3	-0.895435939	4.14616E-06	A
LSM2	-0.520764998	0.009538191	A
MAK31	-0.813923145	4.30386E-05	A
MIC14	-0.92733695	1.16031E-06	A
MPT5	-0.514247242	0.01624283	A
MRPL32	-0.530001594	0.009524318	A
OAC1	-2.004097236	2.02164E-24	A

PFS1	-1.041574687	0.011968229	A
PHM6	-0.74334756	0.000431188	A
PHO5	-1.158188271	9.12423E-08	A
PHO89	-0.900604926	5.58284E-06	A
POP5	-0.568118801	0.005761562	A
POP8	-0.636261012	0.00095524	A
QDR2	-0.675804621	0.000586599	A
RPC17	-0.577357809	0.004954345	A
RPI1	-0.921866786	1.39643E-05	A
RPL29	-0.815796571	0.000706639	A
RPL41a	-0.97650842	0.000101622	A
RPL41b	-0.9967793	6.38269E-05	A
RUF20	-0.953471455	0.000107536	A
SDH6	-0.523786303	0.039458143	A
SNR128	-0.605257911	0.024834086	A
SNR13	-0.860959407	0.00232404	A
SNR189	-1.61045898	1.13392E-05	A
SNR30	-0.637831611	0.001139118	A
SNR31	-0.651335912	0.000817643	A
SNR32	-0.938597366	0.001230356	A
SNR33	-0.749825553	7.87114E-05	A
SNR34	-0.881007171	0.001563335	A
SNR35	-0.986915482	0.000143543	A
SNR36	-1.044669956	0.000364061	A
SNR42	-0.538154689	0.037410767	A
SNR43	-0.889219947	0.000590526	A
SNR45	-0.790898619	0.007896757	A
SNR46	-0.882482169	0.003088374	A
SNR47	-0.704429287	0.021934458	A
SNR49	-1.138182322	3.60036E-09	A
SNR5	-1.070880276	2.91103E-05	A
SNR53	-1.343323358	0.023119984	A
SNR6	-2.075127247	0.026934196	A
SNR61	-1.106080175	8.16259E-05	A
SNR62	-0.945581707	0.005638757	A
SNR63	-0.572525093	0.002525925	A
SNR65	-1.139978391	0.001649645	A
SNR80	-0.823902143	0.038077977	A
SNR83	-0.798965028	2.66289E-05	A
SNR86	-0.546164182	0.012238019	A
SNR9	-0.840059992	0.00020938	A
SPG3	-0.699844002	0.002992471	A
SPL2	-0.743571382	0.000198302	A
SSA4	-0.734221932	0.000359819	A
SUF5	-1.097847323	0.026307309	A
SUP11	-1.734090329	0.008075364	A
SUS1	-0.587887305	0.002038175	A
tD(GUC)O	-1.13899943	0.000516776	A
tF(GAA)P1	-0.831901322	0.018094948	A
TFB5	-0.780432665	5.2157E-05	A
tG(GCC)F2	-2.020679463	0.012560943	A
tG(GCC)G2	-1.449522762	0.036183017	A
tH(GUG)E2	-1.241666434	0.006236041	A

tH(GUG)G1	-1.067222049	0.033079018	A
tH(GUG)M	-1.577627587	0.025556684	A
TIM8	-0.601945705	0.002368969	A
tK(CUU)M	-1.026918154	0.032859974	A
tK(UUU)D	-1.212520645	1.58474E-05	A
tK(UUU)P	-1.018096302	0.001846434	A
tL(CAA)N	-1.183503408	0.026372222	A
TMA19	-0.579317379	0.016993352	A
TMA7	-0.698488561	0.001787629	A
tP(UGG)O3	-0.901233343	0.047161819	A
TPO4	-0.577537781	0.007123898	A
tR(UCU)G3	-1.052584678	0.020155118	A
TSC3	-0.676284574	0.001132717	A
tV(AAC)G2	-1.168236651	0.018737015	A
tV(AAC)L	-0.986122241	0.036052643	A
tV(AAC)O	-0.750816501	0.038751943	A
Unknown	-0.510863739	0.047726623	A
WIP1	-0.558376049	0.006027617	A
YBL029W	-0.925002488	3.94001E-06	A
YBR064W	-3.812092841	0.012395822	A
YCR018C-A	-0.877980777	0.004686086	A
YCR102W-A	-1.377690017	0.041345238	A
YDL085C-A	-0.671367816	0.000426631	A
YDL121C	-0.614059603	0.002350845	A
YDR278C	-1.992665087	0.036256236	A
YER091C-A	-1.54461253	0.042790037	A
YER138W-A	-1.189843107	2.89352E-09	A
YGR022C	-2.075127247	0.040278939	A
YGR174W-A	-0.81722308	0.005579671	A
YJL133C-A	-0.757645057	0.002364603	A
YJR011C	-0.92545296	1.07116E-05	A
YLR346C	-0.553111091	0.041058949	A
YLR428C	-2.190604464	0.04460873	A
YLR456W	-0.615311063	0.001953191	A
YML007C-A	-1.034485262	0.030732546	A
YMR085W	-0.582652165	0.020805355	A
YMR290W-A	-2.050879701	0.000946558	A
YNL109W	-1.06602504	0.041152171	A
YNL211C	-0.663213141	0.006815338	A
YNR075C-A	-0.827199733	0.024023734	A
YOL038C-A	-0.798119899	0.00155873	A
YOR225W	-1.364633864	0.032490107	A
YOR314W	-1.245052248	0.035831548	A
YPL225W	-0.503229704	0.016936135	A
ZEO1	-0.702709841	0.005808924	A
ARO9	-1.132967918	0.019843195	B
BAP2	-0.859375381	0.02732046	B
ILV2	-0.641585459	0.04818287	B
ILV3	-0.988520585	0.020408715	B
ILV5	-0.956956141	0.021812814	B
ISU2	-0.934657685	0.024900662	B
LEU4	-0.80361025	0.031535497	B
PHO8	-0.869847459	0.027115302	B

snR8	-1.172352598	0.028079331	B
YBL029C-A	-0.685684794	0.050393906	B
YKL075C	-11.33667083	0.030206337	B
ADH5	-0.91371471	0.026956535	B
ARG4	-0.641630383	0.049900712	B
ARG7	-0.650009152	0.04934722	B
ARN1	-0.901899438	0.027008687	B
CPA2	-0.926243179	0.024141654	B
DRE2	-0.697804801	0.0428953	B
ENB1	-0.649001298	0.047693068	B
HIS4	-0.646604422	0.047570807	B
LEU9	-0.277276859	0.195628342	B
MAE1	-0.964816095	0.021622047	B
PDR15	-0.852804039	0.029268402	B
PDR5	-0.656679405	0.045631117	B
PHO81	-0.917420978	0.025670539	B
PHO86	-0.662221282	0.046564027	B
RDT1	-0.879633238	0.041604572	B
SCR1	-0.730694306	0.041530274	B
SIT1	-0.777667489	0.034724737	B
snR37	-0.789126897	0.047740909	B
YGL117W	-0.78387062	0.045110355	B

8

Figure 2.1:
        Panels a-c: CNS and EJG equally contributed.
        Panels d-e: CNS and CAT equally contributed
        Panels g-j: CNS
Figure 2.2:
        Panels a: CNS.
        Panel b: TGL
        Panel c: RHS
        Panels d-k: CNS
Figure 2.3:
        Panels a-h: CNS
Figure 2.4:
        Panels a-h: CNS and CAT equally contributed
Supplemental Figure 2.1: panels a-c: CNS
Supplemental Figure 2.2:
        Panels a,b: TGL
        Panels c-e: CNS
Supplemental Figure 2.3
        Panels a-c: CNS
Supplemental Figure 2.4
        Panels a,b: CNS and CAT equally contributed
        Panels c-f: CNS
Supplemental Table 2.1; CNS, CAT, EJG, RHS, all contributed strains to this project
Supplemental Table 2.2: CNS, CAT, EJG, RHS, all contributed primers to this project
Supplemental Table 2.3: CNS and RHS

Chapter 3: Underlying mechanism of the age-related decline in actin cytoskeleton function in of *Saccharomyces cerevisiae*

1   Summary

Aging affects all living organisms and is characterized by the progressive loss of cellular homeostasis, which leads to cellular dysfunction and ultimately to death. Many organelles and subcellular structures, including the actin cytoskeleton, develop age-associated defects. For example, age-associated declines in wound healing have been linked to defects in the actin cytoskeleton which compromise migration of mesenchymal stem cells (MSC) to sites of tissue. Our previous studies revealed that the integrity of actin cables, bundles of F-actin that serve as tracks for cargo movement that are essential for daughter cell formation during cell division, declines with age and that this process affects lifespan. However, the mechanisms underlying aging of the actin cytoskeleton are not well understood. Here, we use the budding yeast *Saccharomyces cerevisiae,* to study how actin dynamics change with age. We find that actin cable assembly is preserved throughout the aging process; however, bundling of F-actin with actin cables declines with age. We observe an age-linked decline in protein levels of Bnr1p, a formin that is also a bundling protein. Moreover, deletion of the yeast fimbrin homolog, *SAC6,* results in loss of actin cable integrity and reduced replicative lifespan. Our studies support a model whereby an age-linked decline in the bundling of F-actin within actin cables contributes to the aging process through effects on the stability and function of these structures in yeast.

2   Introduction

The aging process affects all organisms, causing a decline in cellular function. Mitochondria are an aging determinant through their functions in central metabolism, oxidative stress, and signal transduction. In yeast, mitochondria that are more reduced, have less reactive oxygen species (ROS), and higher membrane potential are preferentially inherited by yeast daughter cells [71, 245]. This process affects cell fitness and lifespan [70, 246]. Interestingly, mitochondria are asymmetrically inherited during stem cell division, and this process affects cell fate [72].

Other studies indicate that the actin cytoskeleton undergoes an age-linked decline which may contribute to the deterioration of bone, skin, liver in advanced age. For example, defects in retrograde actin flow in the leading edge of cells occur in mesenchymal stem cells (MSCs) or T-cells as they age. This, in turn, affects MSC migration to sites of injury, which affects wound healing, and the formation of an immunological synapse between T-cells and antigen-presenting cells, which contribute to immune senescence [211–213, 215, 216].

In yeast, actin cables, bundles of F-actin that align along the mother-bud axis, are essential for cell division through their function in the movement of cellular constituents from mother cells to developing buds. Actin cable also affect lifespan by promoting movement and inheritance of higher functioning mitochondria from mother to daughter cells. Since actin cables undergo retrograde flow, assembly- and motor-driven movement from their site of assembly in the bud toward mother cells, mitochondria must overcome the opposing force of retrograde actin cable flow (RACF) and move from mother to bud on actin cables [56, 70, 71]. Thus, actin cables serve as filters to prevent low function mitochondria from moving into and being inherited by yeast daughter cells.

Our understanding of the mechanisms underlying actin cytoskeleton aging is incomplete. However, recent studies from our lab and others indicate that F-actin stability declines with age, promoting lifespan or healthspan in *C. elegan* and budding yeast (Chapter 2) [217, 218]. Using yeast as a model system, we find that defects in bundling of F-actin within actin cables decline with age and contributes to aging the actin cytoskeleton.

Bundling of F-actin within cables is mediated by 3 proteins: fimbrin (Sac6p), Abp140p, and the formin, Bnr1p. Sac6p is the yeast homolog of the actin bundling protein fimbrin and a member of the calponin homology domain superfamily of F-actin bundling proteins [247]. Sac6p localizes to actin cables and is required for the integrity of those structured [4]. Finally, Sac6p function is regulated by metaphase Cdk1 and undergoes relocalization in response to DNA replication stress [248, 249]. Abp140p is an actin bundling protein that localizes exclusively to actin cables [250]. Bni1p and Bnr1p are the two formin proteins of yeast. Although these proteins have some overlapping genetic interactions, they have distinct activities, localization and regulatory mechanisms. Bni1p and Bnr1p localize the bud tip and bud neck, respectively. They also have different activities. Bnr1p, but not Bni1p, has F-actin bundling activity. Bnr1p also has a 10-

fold higher actin nucleation activity compared to Bni1p [4, 251]. We find that actin cables undergo an age-linked decline in bundling, identifying a role for conserved actin bundling proteins in aging in actin cable and lifespan control in yeast.

3   Results

3.1   Actin cable assembly is preserved during aging

Actin cables undergo a decrease in apparent thickness and an increase in abundance with age (Chapter 2). These results suggest defects in actin cytoskeleton assembly: polymerization and actin cable dynamics. To test this hypothesis, we investigated whether there is a change in the expression or localization of an actin formin protein, Bni1p. Bni1p is a formin that localizes to the bud tip where it facilitates actin polymerization during actin cable assembly. We tagged Bni1p at its chromosomal locus with GFPEnvy, isolated cells as a function of replicative age isolated as described previously and used wide-field imaging to evaluate the localization of Bnip1-GFPEnvy in young and old cells [157]. We find that the localization of Bni1p to the bud tip remains constant throughout the replicative aging process with no changes in Bni1p expression (Fig. 3.1a). Bni1p is a fomrin that drives actin polymerization, which is a force that drives retrograde actin cable flow (RACF). Thus, observing no age-associated defects in Bni1p indicates RACF may also exhibit no defects during aging.

RACF can be by tagging a bundling protein, Abp140p, who exclusively localizes on actin cables. We tagged Abp140p at its chromosomal locus with GFPEnvy and monitored RACF by time-lapse imaging and measure the rates of RACF in living young and old cells. Our initial findings indicate that there are no significant differences in RACF rates between young and old cells (Fig. 3.1b). This suggests that the forces for RACF (polymerization and assembly of actin cables and pulling forces by myosin motor proteins) are not compromised during the aging process.

3.2   Age-associated decline in actin cable bundling

Our previous studies revealed a decrease in actin cable width and an increase in actin cable abundance with age in yeast. Since cables are bundles of actin filaments, we hypothesized that actin cables may unbundle with age. Therefore, we studied how the formin and actin bundling protein Bnr1p, changes with age. We find that Bnr1p expression, based on its localization to the bud neck, declines with age. Bnr1p is expressed at the bud neck in 100% of the young cells analyzed. However, Bnr1p is not detected at the bud neck in 35% of yeast at an advanced age (Fig. 3.2a,b).

In yeast, there are two other actin bunding proteins: Sac6p and Abp140p. Deletion of *ABP140* has no effect on cell growth rates. In contrast, *sac6Δ* cells displayed a significant and profound reduction in their growth rate (Fig. 3.2c). Therefore, we focused our attention on Sac6p. We evaluated actin morphology in mid-log phase in fixed *sac6Δ* cells using AlexaFluor 488-labeled phalloidin (Fig. 3.2d). Deleting *SAC6* results in the loss of the majority of actin cables. It also results in an increase in actin patches in mother cells, which indicates a loss of actin polarity.

Next, we tested whether deletion of *SAC6* affects replicative lifespan in yeast. Our initial findings indicate that cells lacking *SAC6* exhibit a significant reduction in their replicative lifespan compared to wild-type cells (Fig. 3.2e). On average, wild-type cells underwent 23 generations of division, while cells deleted for *SAC6* reproduced for an average of only 7 generations (Fig. 3.2e). Thus, actin bundling activity is critical for cell fitness and lifespan in yeast.

4 Discussion

Recent studies provide evidence that a decline in actin form and function is a hallmark of aging. However, the mechanism of these age-associated changes remains elusive. Aging yeast display altered actin morphology and polarity: actin cables are more abundant, thinner, and depolarized in yeast as they age (Chapter 2). Here, we studied the nature and causes of the age-linked decline in actin cytoskeletal function.

First, we find that RACF and the localization of a formin (Bni1p) that drives actin cable assembly are preserved in aging cells. On the other hand, we find that Bnr1p, a formin protein that also has actin bunding activity, expression declines with age. Previous studies indicate that yeast that contains Bnr1p as the sole formin protein exhibit reduced lifespan compared to wild-type cells [252]. Thus, the decline in lifespan observed when Bnr1p is the sole formin may be due to age-associated expression of Bnr1p.

Since RACF does not change with age and Bnr1p exhibits an age-linked decline in expression, it may not affect RACF. On the other hand, Bnr1p has actin bundling activity. Therefore, Bnr1p may affect aging actin cables through its function as an actin bundling protein. We then tested whether other actin bundling proteins impact the aging process. We find that deletion of the fimbrin homolog *SAC6* led to growth defects, profound effects on actin cytoskeleton morphology, and a dramatically reduced replicative lifespan compared to wild-type cells. Interestingly, another bundling protein, Abp140p, did not change significantly in abundance with age, and deletion of the *ABP140* gene did not alter growth rate.

Age-induced unbundling of actin cables may also contribute to the other age-associated actin phenotypes, such as decreases actin cable width and increased cable number. Intuitively, unbundling cables into individual actin filaments could produce more numerous and thinner actin structures. However, it remains to be seen whether actin bundling proteins directly contribute to these aging phenotypes.

In summary, our evidence supports a model in which actin cables become unbundled in aging cells, resulting in decreased width and increased number of cables (Fig. 3.3). Interestingly, our previous findings also show a decline in actin cable stability with age. It remains to be determined whether there is a link between these two phenotypes; it is possible that declining actin stability in an aging cell initiates the unbundling of actin cables leading to the downstream observed phenotypes.

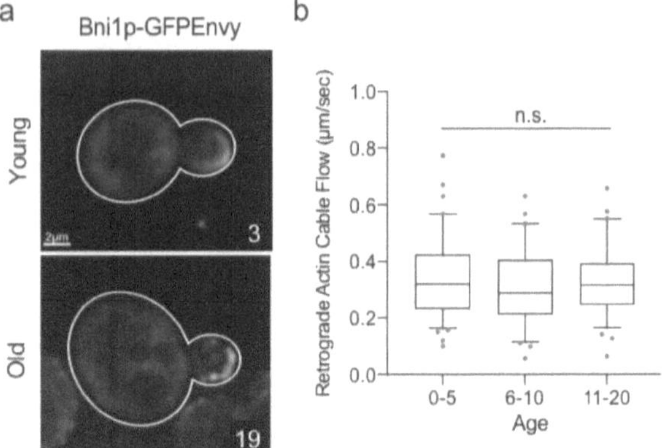

Figure 3.1: Actin cable assembly is preserved during the aging process.
(**a**) Representative wide-field images of Bni1p-GFPEnvy in isolated young and old cells. Number in lower right indicates replicative age of the cell shown, as determined by Calcofluor white staining. Scale bar: 2µm. (**b**) The velocity of RACF in young, middle-aged, and old cells. (n = 40-70 cells). n.s (no significance) based on Mann-Whitney test.

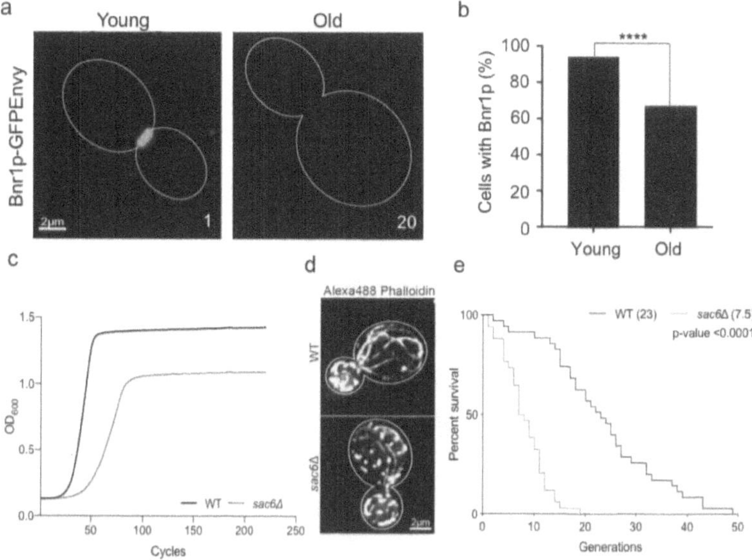

Figure 3.2: Age-linked decline in actin bundling.
(a) Representative wide-field images of Bnr1p-GFPEnvy in isolated young and old cells. (b) Quantification of Bnr1p expression in young and old cells. ****p<0.0001 (Mann-Whitney test). (c) Growth curves of WT and *sac6Δ* cells, showing optical density measurement at 600nm every 20 minutes for 3 days. (d) Representative wide-field images of AlexaFluor 488 phalloidin-labeled actin in WT and *sac6Δ* cells. Scale bar: 2μm. (e) Replicative lifespan of WT and *sac6Δ* cells determined by manual micromanipulation. p <0.0001 (Mantel-Cox test).

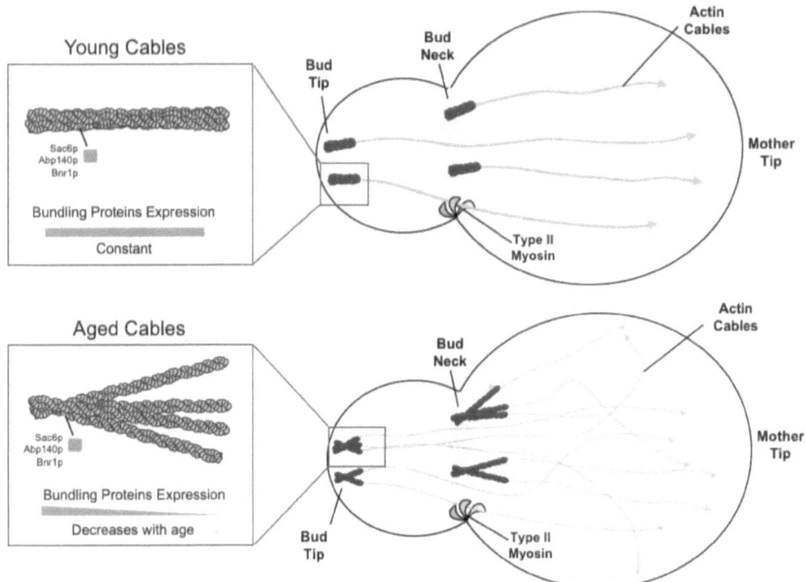

Figure 3.3: The functional model of actin cytoskeleton aging biology.
Schematic depicting a proposed model for age-associated changes in the actin cytoskeleton and the underlying mechanism behind the deterioration of actin.

6   Materials and Methods

Yeast growth conditions

All *S. cerevisiae* strains used in this study are derivatives of the wild-type BY4741 strain (*MATa his3Δ1 leu2Δ0 met15Δ0 ura3Δ0*) or S288C strain (*MATα SUC2 gal2 mal2 mel flo1 hap1 ho bio1 bio6*) from Open Biosystems (Hunstville, AL). All cultures used in imaging and RLS methods were grown to mid-log phase (OD$_{600}$ 0.1-0.5) by shaking at 30°C or 25°C in either rich glucose-based medium, yeast-peptone-dextrose (YPD) or a nitrogen based medium, synthetic complete (SC) medium.

Yeast strain construction

Knockout strains (Supplemental Table 3.1) were created using homologous recombination. Genes of interests were replaced with either auxotrophic marker or drug marker cassettes using appropriate primers. Auxotrophic marker cassettes used were *URA3* and *LEU2*. Drug cassettes used were *kanMX6* or *hphMX6* from pFA6-KanMX6 and PCY3090-02, respectively. For example, CSY036 was created by replacing genomic *YKL075C* with KanMX6. A PCR fragment containing regions homologous to sequences directly upstream and downstream of the start and stop codons of *YKL075C* and coding regions for KanMX6 was amplified from plasmid pFA6-KanMX6 (Addgene, Cambridge, MA) using primers listed in Supplemental Table 2. Cells then were transformed with PCR product using the lithium acetate method and transformants were selected on appropriate selective media plates with 200μg ml^{-1} Geneticin (Sigma-Aldrich, St Louis, MO) to select for KanMX6-positive cells. Positive cells were confirmed by PCR sequencing.

The same techniques as above were used to add fluorescent protein tags to proteins at their chromosomal loci using PCR fragments amplified from plasmids: pFA6a-GFP(S65ST)-KanMX6 and pFA6a-GFPEnvy-KanMX6 (Addgene). Selection for positive transformants was carried out as described above and visually confirmed by wide-field fluorescence microscopy.

Microscopy

Wide-field fluorescence microscopy was performed on an Axioskop 2 microscope equipped with a 100x/1.4 Plan-Apochromat objective (Zeiss, Thornwood, NY) and an Orca ER cooled CCD camera (Hamamatsu). Fluorescence was excited by a pE-4000 LED illumination system (cooLED, Andover, UK). The system was controlled by NIS Elements 4.60 software (Nikon, Mellville, NY).

Visualization of F-actin Cytoskeleton and bud scars

Cells at mid-log phase were concentrated to an OD_{600} of 0.5 and resuspended in 1ml of growth medium. Cells were fixed in 3.7% paraformaldehyde added directly to 1ml of cells at 30°C with shaking for 50 min. Fixed cells were washed three times with 1xPBS, followed by one wash with 1xPBT (PBS containing 1% w/v BSA, 0.1% v/v Triton X-100, 0.1% w/v sodium azide), and actin was stained with 2.5µM AlexaFluor 488-phalloidin (Thermo Scientific, Grand Island, NY) for 25 min at room temperature in the dark. Cells were then washed three times with 1xPBS, and resuspended in Sigma prolong anti-fade mounting solution.

Visualization of stained actin cytoskeleton was carried out by wide-field fluorescence microscopy. Stained cells were mounted in 1.8µl SlowFade diamond antifade mountant (Thermo Scientific, Grand Island, NY) on a glass slide with a #1.5 coverslip. AlexaFluor 488 phalloidin was excited with a 470nm LED and an ET470/40x filter (Chroma, Bellows Falls, VT). Z stacks were captured at 0.3µm intervals using 1x1 binning, 200-300ms exposure time, and gain of 116. Images were deconvolved using a constrained iterative restoration algorithm assuming 507nm emission wavelength with a 100% confidence limit and 60 iterations, using Volocity 5.5 (Quorum Technologies, Guelph, Ontario, Canada). All images were contrast-enhanced with similar parameters.

Bud scars were stained with 25µM calcofluor white (CFW) (Thermo Scientific, Grand Island, NY). CFW was excited with a 407 nm LED. Z stacks were captured at 0.125µm intervals using 1x1 binning, 50ms exposure time, and gain of 116. All images were contrast-enhanced with similar parameters.

Visualization of anterograde mitochondrial and RACF velocity

Cells were grown to mid-log phase in either SC media for anterograde mitochondrial velocity or lactate for measuring RACF. 1.8µL of live cell suspension was spread over the surface of a glass slide and covered with a #1.5 coverslip. The cells were imaged within 5 min after slide preparation. GFPEnvy was excited with a 470 nm LED at 45% power and an ET470/40x filter (Chroma, Bellows Falls, VT). Images were collected at a single focal plane: the middle plane for mitochondria, and 0.5-1µm above the center of the mother cell for actin. Time-lapse images were collected at 500msec intervals for a total of 30 sec using 1x1 binning, 200 ms exposure, and gain of 116. Anterograde mitochondrial velocity was determined by tracking the position of the tip of a moving or elongating tubular mitochondria as a function of time. Mitochondrial movement was scored when there were 3 or more consecutive movements in the same direction. The velocity of retrograde actin cable flow was determined by measuring the change in position of the tip of moving or elongating cables, or the movement of bright fiduciary marks along actin cables, as a function of time.

Growth rates

Growth curves were determined using an automated plate reader (Tecan; Infinite M200, Research Triangle Park, NC). Each strain was grown to mid-log phase in rich, glucose- based media (YPD) and diluted to an OD600 of 0.07 (2.0×10^6 cells/ml). 10 µl of a diluted cell suspension was added to a well containing 200 µl YPD in a 96-well plate. Cells were propagated at 30°C without shaking, and optical density at 600 nm (OD600) was measured every 20 min for 72 hrs. For each strain, three independent colonies were tested in quadruplicate. The maximum growth rate was calculated using the greatest change in OD_{600} over a 240-min interval in 72 hrs.

RLS Analysis

RLS measurements were performed as previously described in [153], without alpha-factor synchronization. Selected strains stored at -80°C were streaked out on YPD plates and grown for 2-days at 30°C. A patch of colonies were selected and grown overnight in liquid YPD at 30°C to mid-log, exponential, growth phase. 10µl of cell suspension was streaked onto a YPD plate. Small-budded cells were isolated and arranged in a matrix using a micromanipulator mounted on a dissecting microscope (Sporeplay, Singer Instruments, Somerset, UK). Upon a complete budding event, mother cells were removed and discarded to leave only virgin cells. The time and number of subsequent daughter cells produced by each virgin mother cell were recorded until all replication ceased.

Quantification and Statistical Analysis

All data were analyzed for normal distribution with the D'Agostino and Pearson normality test. p-values for simple two-group comparison were determined with a two-tailed Student's t-test for parametric distributions and a Mann-Whitney test for non-parametric data. For multiple group comparisons, p-values were determined by a 1-way ANOVA with Dunnett's or Sidak's test for parametric distributions and a Kruskal-Wallis test with Dunn's post hoc test for non-parametric distributions. Survival curves were statistically compared using a Log-rank (Mantel-Cox) test. GraphPad Prism7/8 (GraphPad Software) was used to conduct all statistical analysis. Bar graphs show the mean and SEM; in box and whisker graphs, the box represents the middle quartile, the midline represents the median and whiskers show the 10th and

90th. For all tests, p-values are classified as follows: $*p < 0.05$; $**p < 0.01$; $*** \; p < 0.001$; $****p < 0.0001$.

7   Supplemental Information

Table 3.1:

Strains	Genotype	Source
CSY063	MATa his3Δ1 leu2Δ0 met15Δ0 ura3Δ0	This Study
CSY143	*MATa his3Δ1 leu2Δ0 met15Δ0 ura3Δ0 sac6Δ::LEU2*	This Study
CSY187	MATa his3Δ1 leu2Δ0 met15Δ0 ura3Δ0 act1-V159N ABP140-ENVY-HIS3	This Study
CSY203	*MATa his3Δ1 leu2Δ0 met15Δ0 ura3Δ0 abp140Δ::LEU2*	This Study
EGS168	*MATa his3Δ1 leu2Δ0 met15Δ0 ura3Δ0 CIT1-Envy-GFP-HIS3*	This Study
EGS249	*MATa his3Δ1 leu2Δ0 met15Δ0 ura3Δ0 BNR1-Envy-HIS3*	This Study
EGS251	*MATa his3Δ1 leu2Δ0 met15Δ0 ura3Δ0 BNI1-Envy-HIS3*	This Study

8

Figure 1: CNS and EJG equally contributed

Figure 2: Panels a, b: CNS. Panel C: CNS and EJG equally contributed

Figure 3: Panels a, b: CNS and EJG equally contributed. Panels c, d, e: CNS

Figure 4: CNS

Supplemental Table 1: CNS and EJG

Chapter 4: Imaging the actin cytoskeleton in live and fixed budding yeast cells

1 Summary

Budding yeasts, *Saccharomyces cerevisiae*, are a great model organism to study the actin cytoskeleton function and organization because it simply expresses only 1 actin isoform, unlike mammals. Past research made exploring cell biology difficult in yeast due to the cell wall and its small rounded shape, however, with improved technology and scientific techniques, researchers can acquire high-resolution imaging and allows for detailed multidimensional analysis of a yeast cell. Given these factors, imaging in yeast has emerged as an important tool for eukaryotic cell biology research, specifically the actin cytoskeleton. This chapter describes many techniques and approaches for visualizing the actin cytoskeleton from different methods to optical imaging parameters such as wide-field or super-resolution fluorescent microscopy.

2 Introduction

Thirty six years ago, the actin cytoskeleton was first visualized in fixed budding yeast cells with fluorescent phallotoxins [253]. Subsequently, fluorescent imaging became a useful tool to study actin containing structures. Many elements of the actin cytoskeleton structure and function are conserved within the eukaryotic lineage. Nevertheless, budding yeast actin cytoskeleton is expressed from a single actin gene, making it a fruitful model organism to study a large array of actin organization and cellular dependent functions. The actin cytoskeleton holds many responsibilities like: maintaining cell polarity, direct cellular division, endocytosis and organelle transport.

In yeast, the actin cytoskeleton is composed of two major structures, which are maintained through cellular division, actin patches and cables. Actin patches are endocytic vesicles covered in actin that visually appear punctuated. Actin covered endocytic vesicles are primarily nucleated by the Arp2/3 complex and includes conserved proteins such as: clathrin and its adaptors, membrane-binding protein, endocytic adaptor, nucleation promoting factors (NPFs), kinases, and actin bundling proteins (e.g. Sac6p/Fimbrin) [254]. Early G1 phase of the cell cycle, actin patches polarized towards the nascent bud site, which re-directs its polarization towards the small growing daughter cell during bud emergence and then towards the bud cortex during isotropic growth [4, 42]. Lastly, when the bud cell reaches a similar diameter to the mother cell, actin patches re-distribute to the bud neck and facilitate cytokinesis [4, 42]. The distinct actin patch polarity can be used as an indicator for functional cell polarization machinery.

Another important structure are actin cables, where they execute multiple roles. Actin cables are constructed by bundling multiple actin filaments together and these cables are elongated through the entire length of the yeast cell. Actin filaments are bundled by conserved bundling proteins: Sac6p (fimbrin), Abp140p, and Bnr1p. Actin cables are further stabilized by two tropomyosin isoforms (Tpm1p and Tpm2p). Polymerization of actin cables begins at pools of actin that are in the bud tip and bud neck. At these sites' actin polymerization begins, where actin barbed (plus ends) exist and then are elongated towards the mother tip. At the highest polarized state during cell division, the actin cables align along the mother daughter axis. However, late stages of cell division reorganize the actin cables to orient their plus ends of both mother and daughter cell towards the neck and initiate cytokinesis. Actin cables are extremely important for yeast because they lay down tracks for trafficking intracellular cargo such as: mitochondria, vacuoles, secretory vesicles and mRNA between the diving cells [255]. Moreover, analogous to mammals, actin cables are highly dynamic where their polymerization occurs in a retrograde manner, daughter to mother cell, while cargo is transported in an anterograde direction, mother to daughter cell [57].

Visualizing actin dynamics can only be captured through living-imaging. This can be achieved by endogenously tagging actin bundling proteins or fluorophore conjugated small peptides. Here, we describe alternative methods for visualizing the actin cytoskeleton using conventional wide-field imaging to create time-lapse imaging and quantify using open-source imaging analysis software, Fiji-ImageJ.

3   Live-cell Imaging

Imaging techniques and technical methods have evolved over time to allow for efficient live-cell imaging of actin dynamics in yeast cells. These techniques involve endogenously tagging a fluorophore protein of genes at their chromosomal loci to visualize actin movement. Here, we describe the steps it takes to visualize actin movement by describing methods for tagging yeast genes at their chromosomal loci, determine functionality of proteins with fluorescent tags and tagged gene products expressed either from endogenous or other promoters. Followed by describing top fluorescent protein tagging methods for actin movement in yeast. Closing this chapter, we discuss imaging methods and provide descriptive steps to carry out quantitative analysis of actin dynamics.

3.1   Detection of cellular structures using targeted fluorescent proteins

Live-imaging of yeasts' actin cytoskeleton can be visualized through fluorescent proteins (FPs) or small amino peptide conjugated with FPs. The commonly used fluorophore is GFP and its variation, however, other FPs can be used. Here, we describe what criteria one should follow when exploring the use of different FPs and methods of how to endogenously tag a protein at the chromosomal locus.

3.1.1   Choosing a fluorescent protein

The green fluorescent protein (GFP) revolutionized live-cell imaging in cell biology. GFP was discovered and cloned from the jellyfish Aequorea Victoria. Currently, GFP has a growing array of FPs with ranging colors and molecular properties. Moreover, combining laboratory mutagenesis with novel FPs, such as mCherry from Discosoma sp. coral and Teal from Clavularia sp. coral, yield variety of colored FPs with improved brightness, faster folding, and decreased oligomerization [256].

To perform a successful live-cell experiment requires a photostable FP to withstand repeated imaging and light exposure. In addition, emission of FP must be bright enough to illuminate structure of interest to over power background fluorescence and detector noise, while being spectrally distinct from other labels being used. FP brightness is determined by its intrinsic brightness (the product of extinction coefficient and quantum yield), and the behavior of the FP upon ectopic expression (rate of folding and stability). The amount of signal that is excited and detected depends on imaging system properties (light sources, lenses, filters and detectors). For more information on selecting fluorescent proteins, Tally Lambert and Kurt Thorn prepared an extensive database of physical properties of FPs, available at http://nic.ucsf.edu/FPvisualization/. Lastly, the FP must not be toxic or disrupt the cellular behavior or protein function to which it is fused.

3.1.2   Tagging endogenous proteins

Homologous recombination methods are used to chromosomally tag a protein of interest. This involves the use of a double-stranded linear DNA that encodes the tag of interest and a selection marker, which is then homologously recombined and inserted into the target site. The double-stranded linear DNA with the tag of interest and selection marker is known as the insertion cassette and insertion cassettes are produced by PCR using a tagging vector as a template. Templates for tagging vectors encompass variety of FPs (e.g. GFP, mCherry, mCitrine), epitopes (e.g. HA, myc), and affinity tags (e.g. GST, TAP, His) that are linked with a large array of selection markers conferring drug resistance or auxotrophic rescues. Families of tagging vectors have been constructed to share PCR-priming sequences. Therefore, a single set of primers made from one template family can be used insert different tags associated with that template for a given target gene. Additionally, there are template vectors that are available with different expression level control of tagged genes from endogenous promoters, constitutively active promoters (e.g. AHD1, GPD1), and regulatable promoters (e.g. GAL1). Moreover, FP and epitope tags can also be used for biochemical techniques including: affinity purification, immunoprecipitation, western blot analysis, and immunofluorescence.

Such cassette families, like the pOM family, permit the removal of the selectable marker after tagging [257]. These cassettes are flanked by LoxP sites and can be removed by bacteriophage Cre recombinase that is conditionally expressed from a plasmid [258]. Removing the selectable marker is useful in several situations: 1) create an N-terminal tag with no separation between the tag and the endogenous promoter of the tagged protein; 2) for inserting a tag internally within the coding region of the gene of interest; 3) for multiple rounds of tagging at the same locus; and 4) for use in yeast strains with a limited number of selectable markers. Commonly available tagging vectors are listed in Table 4.1.

Table 4.1: Yeast tagging cassette vectors

Plasmid family	Tag position	Promoter	Tags	Markers
pFA6a[1]	C terminal	endogenous	GFP(S65T) 3xHA 13xMyc GST GFPEnvy[2] GFPIvy[2] GFPγ[2]	*TRP1* *kanMX6* *HIS3MX6*
pFA6a-PGAL1[1]	N terminal or internal	*GAL1*	GFP(S65T) 3xHA GST	*TRP1* *kanMX6* *HIS3MX6*
pUR[3]	C terminal	endogenous	DsRed	*HIS3* *URA3(K.l.)*
pYM[4]	C terminal	endogenous	yEGFP EGFP EBFP ECFP EYFP DsRed, DsRedI RedStar, RedStar2 eqFP611 FlAsH 1xHA, 3xHA, 6xHA 3xMyc, 9xMyc 1xMyc+7xHis TAP	*kanMX4* *hphNT1* *natNT2* *HIS3MX6* *klTRP1*

			Protein A	
pKT[5]	C terminal	endogenous	yEGFP	*KanMX*
			yECFP	*SpHIS5*
			yEVenus	*CaURA3*
			yECitrine	
			yESapphire	
			yEmCFP[6]	
			yEmCitrine	
			tdimer2[7]	
			yECitrine+3xHA	
			yECitrine+13xMyc	
			yECFP+3xHA	
			yECFP+13xMyc	
pOM[8]	N terminal or internal	endogenous[9]	yEGFP	*kanMX6*
			6xHA	*URA3(K.l.)*
			9xMyc	*LEU2(K.l.)*
			Protein A	
			TEV-ProteinA	
			TEV-GST-6xHis	
			TEV-ProteinA-7xHis	
pCY[10]	C-terminal	endogenous	Cerulean	*HygromycinB*
			yECFP	*Zeocin*
			yEmCFP	
			yEGFP	
			Venus	
			yEVenus	
			yECitrine	
			yEmCitrine	

			mCherry	
			mEos2	
			yEc-MYC	
			yEHA	
			yEFlAsH	

[1] (Longtine *et al.*, 1998)[259]
[2] (Slubowski *et al.*, 2015)[260]
[3] (Rodrigues *et al.*, 2001)[261]
[4] (Janke *et al.*, 2004)[262]
[5] (Sheff & Thorn, 2004)[263]
[6] monomeric version
[7] tandem dimer of DsRed
[8] (Gauss *et al.*, 2005)[257]
[9] After Cre-mediated removal of auxotrophic marker
[10] (Young *et al.*, 2012)[264]

Tagging vector primers should be designed with sequences to hybridize with both the tagging vector and the target site for homologous recombination within the yeast chromosome. Then the insertion cassette is produced by PCR using the tagging vector as a template. The resulting amplified DNA is transformed into yeast using a standard protocol [265]. Then to select recombinants that carry the inserted tag use the associated selection marker of the insertion cassette, followed by PCR-based screening to ensure correct site insertion of the homologous recombination through PCR fragment basepair size comparison.

3.2   Visualization of the actin cytoskeleton dynamics

Actin patches and actin cables are two major F-actin-containing structures that remain through the budding yeast cell cycle. Actin patches are endosomal vesicles coated with F-actin filaments and their localization changes with cell cycle progression from: budding site, cell cortex, and bud neck. Actin cables are bundles of F-actin filaments that are aligned along the mother-bud axis and are highly dynamic, where the continuous actin polymerization flow occurs from bud-to-mother direction, creating a retrograde actin cable flow (RACF) [57]. These cables are important for transporting cargo between the mother and daughter cells. Such cargo that is trafficked along actin cable are: secretory vesicles, mRNA, spindle alignment elements, mitochondria, Golgi and vacuoles. Regulation of actin cable dynamics plays a major role in regulating mitochondria quality [70]. Therefore, studying actin cable dynamics is important for understanding a yeast's cellular biology. Two major methods used to capture actin dynamics are either endogenously tagging a fluorophore to conserved actin binding proteins or to a small peptide. Typically, the major actin-binding protein used to visualize is Abp140p. Here, we will describe the two main visualization methods we use. Table 4.2 lists other known and functional FP labeled actin-binding proteins that serve as probes for actin patches and actin cables in budding yeast.

Table 4.2: GFP-tagged actin proteins in budding yeast

Actin-containing structure	GFP tagged proteins
Actin patch (early)	Las17p [266, 267], Sla1p [267, 268], Pan1p [267, 269], End3p [270], Sla2p [267, 271], Bzz1p [272], Vrp1 [273]
Actin patch (mid)	Myo3p [274], Myo5p [275], Bbc1p [272],
Actin patch (late)	Abp1p [20, 267, 276, 277], Arp2p [266], Arp3p [266], Arc15p [267, 278], Sac6p [20, 279], Cap1p [280], Cap2p [281], Scp1p [282], Cof1p [283]
Actin cable (vegetative)	Abp140p [57]
Actin cable (G_0)	Lifeact (Abp140 aa 1-17) [284]

3.2.1 Actin dynamics visualization with fluorescent tag actin bundling protein, Abp140p

In earlier research, tagging yeast actin-encoding ACT1 gene with GFP compromised the protein function. Moreover, plasmid ACT1-GFP born expression did not rescue the loss of the endogenous ACT1 gene [276]. The lack of rescue and defective function is presumably a result of GFP tagging because every surface of the actin protein is involved in protein-protein interactions from nucleation to polymerization, motor proteins and other force generators interactions, microfilament assembly, capping, cross-linking, and severing. As an alternative, actin-binding proteins are more susceptible substituted target for FP tagging and are often more tolerant of tagging with FPs. For example, tagging actin-binding proteins like Abp140p and Abp1p with C-terminal GFP produces a fluorescent signal that localizes to actin cables and actin patches, respectively (Fig. 4.1) [57, 58]. Other actin proteins tagged with GFP with known preserved function are listed on Table 4.2. However, GFP fluorescent tagging of Abp140p emits a low signal, which can be improved when culturing cells in a non-fermentable carbon sources, such as lactate. Cells grown in lactate medium express cables that are thicker, henceforth, more visible. Alternatively, improved signal to noise ratio can be achieved by endogenously tagging actin-binding proteins with GFPEnvy. GFPEnvy is an improved version of GFP with brighter signal and photostable, which was constructed by combining mutations found in Superfolder GFP and GFPγ [260]. Tagging with GFPEnvy improves signal intensity and photobleaching, which occur during time-lapse video capture. Nevertheless, not all fluorophores are created equal. GFP and GFPEnvy do not disrupt actin dynamics, however, other fluorophore proteins do, like td-Tomato. Thus, it is important to test whether actin dynamics is preserved when using other fluorescent proteins (FP). Actin dynamics captured by live-cell imaging can permit the analysis of RACF velocity, along with, visualization of cargo trafficking, such as the movement of mitochondria along actin cables (Fig. 4.1) [56].

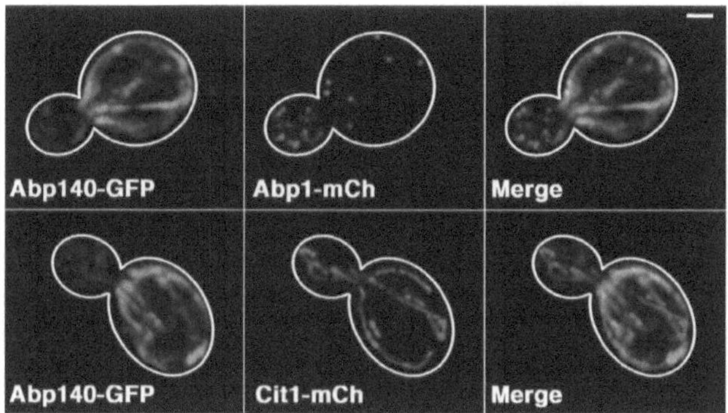

Figure 4.1: Colocalization of the actin cytoskeleton with cargo. Yeast cells were grown in YPG and washed in SC medium prior to imaging. **(Left panels)**: Actin cables were visualized with Abp140-GFP. **(Middle panels)**: Two cellular cargo structures, actin patches (top) and mitochondria (bottom), were visualized by tagging Abp1p and Cit1p with mCherry, respectively. Images were captured using standard GFP filters and 300 ms exposure time for GFP and standard DsRed filters and 200 ms exposure time for mCherry. **(Right panels)**: merged images indicating colocalization of actin patches and mitochondria along the actin cytoskeleton. Scale bar = 1 μm.

3.2.2 Analysis of actin dynamics

The methods described here employ the open source software, Fiji-ImageJ [285] as the tool of choice in our own laboratory. Similar functions are available in most image analyzing and processing software packages.

Tracking movements involves marking the position of an object at successive timepoints. Structures with elongated, irregular or dynamic shapes, like actin cables, a point can be defined on the object of interest and can be used as the reference point for velocity measurements. For example, when tracking actin dynamics with Abp140-GFP-labeled actin, heterogeneity of Abp140 binding produces bright dots along the actin cable, and these areas serve as fiduciary marks to track movement [57]. Tracking can provide several quantitative measurements including velocity, distance traveled, and direction of movement. Here, we show how to calculate velocity.

3.2.3 Measuring retrograde actin cable flow velocity

To make measurements in Fiji-ImageJ:

1. Identify individual cells for actin movement by rapidly scrolling through the movie from beginning to end using the **Point Selection** tool from the Fiji-ImageJ toolbar.

2. Capture screen of identified individual cells through **Plugin > Utilities > Capture screen**.

3. Crop each identified cell into individual videos for tracking actin movement.

4. To track actin cable movement, use Fiji **manual tracking plugin**.

5. Manually import microscope paraments which include: **time interval (30 s), x/y calibration (pixels), and z calibration (z-slice size).**

    a. Additional recommended parameters to visualize the tracked cable. This overlays manual tracing on top the video:

        i. Under tracking subsection, click **show path**.

        ii. Under drawing subsection, click **overlay dots & lines**.

6. To track cables, begin at point of interest:

    a. The polymerizing, plus end.

    b. The minus end, actin cable tip.

    c. Fiduciary marks.

7. Track cables by using Point Selection tool from tool bar. Check **Auto-measure** and **Auto-Next Slice**.

8. At first sight of movement (first timepoint), click **Add track.**

9. Continue marking motile structure for the duration of observable movement.

10. Once tracking is finished, click **End track**. This will produce two results windows.

    a. Manual tracing overlay of actin cables window.

        i. Confirm movement of actin cable match manual tracing.

b. Data log window, which recorded values of each logged data point.

     i. Column 3: number of slices tracked.

     ii. Column 6: total distance of tracking.

11. Calculate velocity

a. $Velocity = \dfrac{Total\ Distance}{Total\ Time\ (Slice\#*Video\ Time\ Interval)}$

To find mean velocity, calculate the average of all incremental velocities.

4   Fixed-cell Imaging

Actin structures can be visualized either by standard immunofluorescence or by binding the small peptide phalloidin. The main difference between these reagents is that phalloidin binds only polymerized, filamentous actin (F-actin), whereas antibodies will bind both F-actin and unpolymerized, globular actin (G-actin). Typically, the polymeric structures are of most interest, so phalloidin is the preferred reagent and has an added advantage of being a simpler staining protocol. However, in cases where visualizing G-actin is important (for example, to assess a change in equilibrium between F- and G-actin), the antibody may be a better choice.

While observations in fixed cells cannot capture the dynamic nature of the actin cytoskeleton, they are very important in acquiring structural data. Enhanced detailed structures can be revealed by using super-resolution structured illumination microscopy (SIM) techniques. SIM provides spatial resolution that surpasses the diffraction limit of conventional fluorescence microscopy. As a result, structures and fine details that would otherwise be undetectable are made evident. Here, we describe methods for visualizing the actin cytoskeleton using both conventional wide-field fluorescence microscopy and SIM through either immunofluorescence or phalloidin staining. We also describe how to quantify actin cables using freely available image analysis software.

4.1   Materials and Methods

The paraformaldehyde fixation detailed here preserves cellular structures by creating covalent hydroxymethylene bridges between spatially adjacent amino acid residues. Other fixatives, such methanol and acetone, are compatible with some immunofluorescence staining, including the actin cytoskeleton.

However, these fixative solutions are not recommended when evaluating additional organelles, like mitochondria, because they cause solubilization of many membranes [278].

Fixed cells can be stained by various methods. For immunofluorescence staining, the cell wall must first be removed with zymolyase. The resulting spheroplasts are then exposed to pre-treated primary and secondary antibodies in a staining chamber and mounted on a polylysine-coated coverslip.

If antibody staining is not required, we recommend leaving the cell wall intact in order to preserve cell integrity and 3D shape. Intact cells can be stained directly with non-antibody agents such as fluorescent phalloidin and then mounted on a slide. For SIM imaging, intact cells can be further immobilized by binding to a coverslip using the lectin Concanavalin A.

4.2   Fixation of yeast cells

1.   Grow a 5 mL liquid culture of cells in a 50-mL conical-bottom tube to mid-log phase (OD_{600} = 0.1-0.5).

2.   Add paraformaldehyde to the culture medium to a final concentration of 3.7%.

3.   Incubate cells under normal growth conditions for 50–60 min.

4.   Concentrate 1 x 10^7 cells by centrifugation (30 s at 10,000 x g).

4.2.1   Visualization of the actin cytoskeleton in fixed cells using fluorescent phalloidin

The following protocol is for staining of cells in suspension. Any fluorophore of choice conjugated to phalloidin can be used like rhodamine phalloidin or AlexaFluor-488 phalloidin. However, we recommend using AlexaFluor488-phalloidin for imaging actin either with wide-field microscopy or super-resolution structure illumination microscopy because AlexaFluor-488 phalloidin has a constant, reproducible, robust, photostable signal.

1.   Grow 5 mL liquid overnight mid-log phase (OD_{600} = 0.1-0.5) in a 50-mL conical bottom tube.

2.   Fix cells with final 3.7% final concentration of 20% paraformaldehyde for 50 min to 1 h at 30°C shaking incubator.

3.   Wash fixed cell pellet 3 times with 500 µL either wash solution or 1xPBS, concentrating each time by centrifugation (30 s at 10,000 x g).

4. Wash cells with 500 µL of 1xPBT and resuspend in 25 µL of PBT.

5. Add 20 µL of AlexaFluor-488 phalloidin to a concentration of 2.5 µM and incubate for 25 min at RT.

6. Wash cells with 100 µL of 1xPBS.

7. Mount cells for imaging. For enhanced immobilization of intact cells, resuspend in 30-60 µL 1xPBS and mount on a ConA-coated coverslip. For a simpler preparation, continue to the next step.

8. Resuspend cells in 7-10 µL of SlowFade mounting solution. Place ~1.5 µL of cell suspension onto a microscope slide. Place a 22x22 mm coverslip on top of the cells and seal the edges of the coverslip with nail polish.

4.2.2 Attachment of intact cells to a coverslip with ConA

Cells can be attached to ConA-coated coverslips either before or after paraformaldehyde fixation (Fig. 4.2).

1. Prepare high-performance coverslips (18-22mm square) for structure illumination microscopy by washing any excess debris with distilled H_2O. Wick away any excess liquid and place on filter paper to air dry.

2. Place clean coverslip on a Parafilm platform and add 20 µL of 2 mg/mL ConA solution on top of the coverslip and spread using the side of a pipet tip.

3. Incubate in staining chamber for 30 min at RT.

4. Remove unbound ConA by rinsing gently with adding distilled H_2O to the corner of the coverslip until the coverslip is full of liquid and then rotate to remove liquid.

5. Allow coverslip to air dry under dark chamber and store in chamber until needed.

6. Place 30 µL of fixed cell suspension on the coverslip and spread using the side of a pipet tip.

7. Incubate in staining chamber for 15 min.

8. Wash off excess cells by gently applying distilled H_2O as in step 4, repeat wash step 3-4 times.

9. Remove excess liquid by touching the edge of the coverslip to filter paper.

10. Place 7.5 µL of mounting solution onto the center of a slide. For optimal signal preservation (which is essential for SIM), we recommend using Prolong Diamond anti-fade mountant for best preservation of signal intensity. Alternatively, Prolong Gold can be used as the mounting solution.

11. Place the coverslip cell side down onto the mounting solution. Apply gentle pressure on the coverslip to push out any air pockets that may have formed.

12. Allow to cure at RT in the dark for 24 hrs.

13. When using Prolong Diamond, cures cells to the microscope slide, thus, there is no requirement to seal the coverslip edges with nail polish.

14. To preserve the signal intensity image as soon as possible. Do not store for extended periods of time as this will diminish the fluorescent signal.

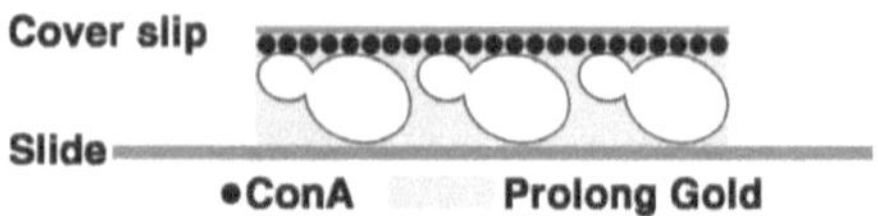

Figure 4.2: Schematic for ConA immobilization of cells onto a coverslip. Coverslips are coated with ConA to immobilize cells. Prolong Gold or other mounting solution is placed onto the slide, and coverslips are placed directly on the mounting medium with cells facing down.

4.3 Imaging methods

Here, we list, in detail, two methods for imaging fixed budding yeast cells: wide-field microscopy with deconvolution; and structured illumination microscopy. Deconvolution is a computational that removes out-of-focus light from wide-field fluorescence images to reveal three-dimensional information [286]. Deconvolved wide-field image provides better signal:noise ratios and more sensitive in comparison to point-scanning confocal imaging and spinning-disk confocal, respectively. The best deconvolved results are images captured with a z-series from 0.2-0.3 µm. Here, we describe two methods to deconvolve wide-field images: algorithms by Volocity (Quorum Technologies) and a free open-source software, Fiji-ImageJ.

Alternatively, high-resolution images can be captured by independent deconvolved imaging techniques such as Structured illumination microscopy (SIM), which allows the resolution to be improved 2-fold the normal wide-field conventional methods [287]. Image quality can be greatly improved when using the SIM, however, it requires a specialized microscope system, additional sample preparation, and increased imaging process time. Nevertheless, suitable SIM samples yields dramatically improved resolution (Fig. 4.3).

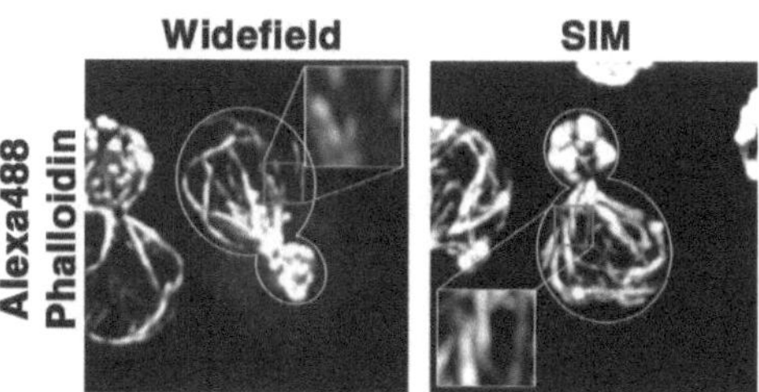

Figure 4.3: Comparison of conventional wide-field microscopy with deconvolution and SIM. Wild-type cells fixed in 3.7% paraformaldehyde, and stained with 2.5 µM Alexa488-phalloidin. **(Left panel)**: Images were collected at 0.3 µm Z-series intervals using conventional wide-field microscopy with: 1x1 binning, 116 gain, 250 ms exposure time, and 470nm LED excitation with standard GFP filter for Alexa488. Deconvolved (Volocity) image using 60 iterations at 100% confidence interval. **(Middle panel)**: Images were collected at 0.125 µm Z-series intervals using Nikon super-resolution structured illumination microscope with: 1x1 binning, 200 gain, 300 ms exposure time, and 488nm laser at 45% power. Image reconstruction was performed using Nikon Elements software with the following parameters: illumination modulation contrast=2, high-resolution noise suppression=1, out of focus blur suppression=0.2. **(Inset)**: 3x magnification of the boxed area for comparison of resolution. Scale bar, 1 µm.

4.3.1 Visualization of fixed cells by wide-field deconvolution

On average, yeast cells are 4-6 µm in diameter, therefore, require a high-magnification objective lens and high-resolution camera. It is important cameras to be able to support low-intensity illumination and short exposure times to minimize photobleaching. Additionally, it is imperative to use filter sets with parameters that maximize light throughput, while preventing bleed-through from multi-color imaging experiments. Captured images then can be imported into Volocity software (Quorum Technologies) or Fiji-Image J for either deconvolution or quantification [285].

Deconvolution protocol using Volocity

1. Optimize signal levels by adjusting fixation and staining conditions, and using as fresh a sample as possible.

2. Set exposure time and camera gain so that pixel values fill as much as possible of the dynamic range of the camera, without saturation.

3. Acquire z-series of images at a focus interval of ~0.2 µm.

4. Load datasets in Volocity and verify that the x, y, and z scale are correctly set. If using open-source, Fiji-Image J, for deconvolution skip to the next section.

5. For each channel, generate a new calculated PSF. When prompted, enter the numerical aperture of the objective lens and the maximum wavelength of fluorescence emission (e.g. 507 nm for AlexaFluor 488-phalloidin).

6. Using the calculated PSF for each channel, perform iterative deconvolution with 60 iterations, 100% confidence limit.

4.3.2 Visualization of cells by structured illumination microscopy (SIM)

SIM imaging improves the resolution by two-fold compared to conventional wide-field microscopy [287]. This resolution reveals complex details like fine structures of the actin cytoskeleton that were not visible using conventional wide-field imaging (Fig. 4.3, Table 4.3). The improved resolution from SIM is generated from modulating the fluorescence excitation pattern by means of a grating. The interaction between the illumination pattern and fine sample details produces a Moiré effect. The super-imposed grating is then rotated at specific angles to generate multiple sets of Moiré patterns, which can be reconstructed to restore details not captured by standard illumination. Importantly, "super-resolution" or "high-resolution" SIM (SR-SIM) differs from previously marketed structured illumination systems such as Apotome, which only provide optical sectioning only. SR-SIM, in contrast, doubles resolution in the x, y, and z dimensions.

Table 4.3: Comparing the number of actin cables and patches between conventional wide-field microscopy and SIM. The number of actin cables and patches in mother cell. Actin cables parallel to the mother-bud axis and greater than half the length of the mother cell were only counted.

Method	Actin cables/cell (± SEM)	Actin patches/cell (± SEM)	n (cells)
Wide-field with deconvolution	8.42 ± 0.21	1.84 ± 0.13	173
SIM	11.37 ± 0.25	2.6 ± 0.20	67

4.3.2.1 Sample preparation and imaging

There are several criteria that must be met for optimal SIM imaging:

1. Minimal movement in the sample. We recommend fixing and immobilizing cells, and using a hard-set mounting medium. In addition, the room and equipment should be stable in temperature and free from drafts.

2. Maximized signal intensity and photostability. Alexa488-phalloidin is preferred for labeling the actin cytoskeleton. Mounting medium should show excellent anti-bleaching performance; Prolong Diamond is recommended.

3. Predictable optical performance. A #1.5 (0.170 mm) coverslip is recommended. Because standard coverslip thickness varies by tens of micrometers, which may induce spherical aberration, high-performance coverslips (manufactured with smaller variation in thickness) should be used. In addition, mounting medium should have a consistent, stable refractive index.

4. Adequate Z sampling. Z-series of images should be acquired at a focus interval of ~0.125 μm and can vary depending on the focus of the experiment.

5. Appropriate pixel values in raw data. Exposure time and camera gain should be set so that pixel values fill no more than 1/4 of the dynamic range of the camera. EM gain for EMCCD cameras should not exceed 150.

4.3.2.2   Reconstruction

Parameters for SIM reconstruction vary with the software used. Here, we describe options for Nikon Elements software (Nikon Instruments, Melville, NY). The following three parameters should be adjusted for each sample. Recommended values for images of actin given.

1. Illumination modulation contrast. Higher values reveal more detail, but may induce artifacts. Recommended values: 0.5–2.0

2. High-resolution noise suppression. Higher values suppress artifacts due to pixel noise, but also suppress fine details. Recommended values: <1.

3. Out-of-focus blur suppression. Higher values suppress out-of-focus signal, essential for densely labeled areas, but also reduce overall signal and resolution. Recommended values: 0.1–0.2.

4.3.3   Quantitative analysis of actin cable number, thickness, and intensity using line profile measurements

Actin cable quantity and structure within a cell vary with age, metabolic state, and other factors, through a host of actin-binding and regulatory molecules. To quantify these effects, we use a line profile analysis method. Since actin cables are organized in parallel of the length of the mother cell, a line profile drawn across the center of the mother cell effectively samples the cable population and can be used to determine the number, thickness, and actin content of actin cables (Fig. 4.4). When the pixel intensity along the line is plotted, cables appear as high-intensity peaks. The number of peaks coordinates the number of cables, and the width of each peak corresponds to the thickness of the actin cable (Table 4.4). The height of the peak or the area under the curve can be used to estimate relative actin content. Although this analysis method is objective, it under estimates the abundance of actin cables. Alternatively, cable quantification can be executed manually with either Volocity or Fiji-ImageJ. Manual methods are more sensitive in capturing cable number, however, it is not as objective as line profile analysis and requires a trained eye to not over estimate cable number. The following protocol describes how to use Fiji-ImageJ [285] to perform these measurements.

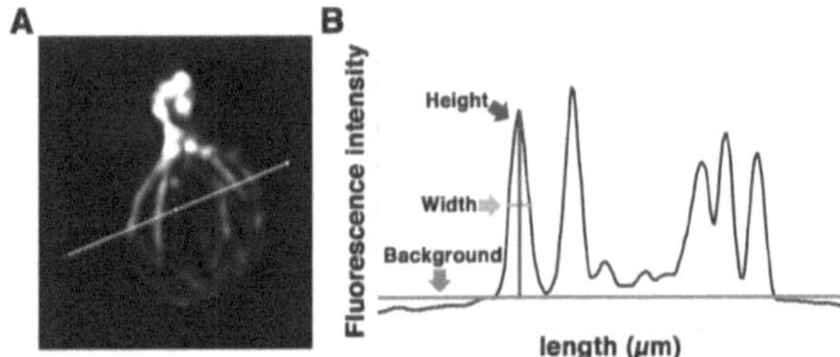

Figure 4.4: Schematic for measurements of actin cable number, thickness, and content using line profiles. (A) Wild-type cells were imaged using Alexa488-phalloidin. A line ROI (yellow) is drawn near the center of the mother cell perpendicular to the mother-bud axis. **(B)** Schematic of line profile measurement of actin cable thickness and actin content. The background level, determined by averaging the intensity of regions outside of the cell, is shown in red. A vertical line is then drawn from the maximum point of the peak to the background line as shown in blue. This represents the maximum intensity of the cable. A horizontal line (green) is then drawn across the peak at half of the maximum intensity. This represents the thickness of the actin cable.

Table 4.4: Line profile measurements reveal carbon source-dependent changes in actin cable thickness. Wild-type cells were grown to mid-log phase in glucose versus lactate medium. Cells were fixed in 3.7% paraformaldehyde and stained with 1.65 µM rhodamine-phalloidin for 35 minutes. Z-series were collected throughout the cell at 0.3 µM intervals through 6 µM using wide-field microscopy using 1x1 binning, 200 ms exposure and a metal halide lamp and standard dsRed filters.

Medium	Actin cable thickness, µM (± SEM)	n (cells)
Glucose	0.241 ± 0.007	83
Lactate	0.288 ± 0.006	123

4.3.3.1 Creating line profiles

1. Collect z-stack images of cells stained with fluorescent phalloidin. It is recommended that images be deconvolved using Volocity or comparable software to remove out-of-focus information and increase the signal:background ratio.

2. Make a maximum-intensity projection of the z stack using **Image > Stacks > Z Project**; choose Projection Type: Max Intensity.

3. If the image has more than one channel, make sure the actin channel is selected.

4. Set up Fiji-ImageJ with the following parameters:

   - **Edit > Options > Input-Output**. Under results table Options, check Copy Row Numbers and leave other options unchecked.

   - **Edit > Options > Profile Plot Options**. Set width and height equal to the width and height of your image. Set Minimum Y to 0 and Maximum Y to the Image Max, obtained from Analyze > Histogram.

   - **Analyze > Set Measurements**. Check Area, Display Label.

   - **Analyze > Set Scale**. Make sure that the spatial scale is set accurately.

5. Make sure the Line ROI tool is set to the Straight Line mode; right-click on the tool to change the mode if needed. Double-click on the Line ROI tool and set Line Width to 3. Do not check Spline Fit.

6. Draw a line ROI perpendicular to the mother-bud axis, near the middle of the mother and avoiding actin patches as shown in Fig. 4A. Both ends of the line should extend into the background outside the cell.

7. **Analyze > Plot Profile**. A Plot window should appear showing Gray Value (pixel value) versus Distance.

8. If desired, save the plot image (**File > Save**). Data can be saved (click **Save** on the Plot window) or copied and pasted into a spreadsheet application such as Excel (Click **Copy** on the Plot window, then paste into the other application). Proceed to the next section to quantify the profile data using ImageJ.

4.3.3.2   Measurements of cable number, thickness, and intensity

1. Count the number of peaks in the profile. This represents the number of actin cables within the mother cell.

2. Determine a background level by inspection or by using a spreadsheet to average the values from the parts of the profile that are outside the cell. In ImageJ, draw a horizontal line across the plot at that gray value, as shown in red in Fig. 4.4.

3. Cable thickness is defined as the width of the peak at half-maximal height above background. To find the height, using the Line ROI tool, draw a vertical line from the maximum height of the peak to the background line, as shown in blue in Fig. 4.4. Holding the shift key while drawing a line forces the line to be either vertical or horizontal.

4. Press T to add the vertical line to ROI Manager. **Analyze > Measure**. The Length value represents the maximum cable intensity.

5. In the ROI manager window, check Show All and Labels. A small number will appear in the middle of the vertical line, marking half the peak height.

6. Draw a horizontal line across the peak at half the peak height as shown in green in Fig. 4.4. **Analyze > Measure.** The length of this horizontal line represents the thickness of the actin cable.

7. Click the Wand tool and select an area under the peak above the background line. **Analyze > Measure**. The **Area** measurement represents integrated density of the cable and can give a relative measurement of actin content for each cable.

Chapter 5: Summary and Perspectives

1 Summary

The aging process is unforgiving and targets all cells, tissues, and organs. Classical aging hallmarks are vast and growing. In this dissertation, I focused on relevant hallmarks that are conserved throughout the eukaryotic kingdom: mitochondrial dysfunction, loss of proteostasis, nutrient sensing dysregulation, and actin cytoskeleton disorganization [2, 160]. The actin cytoskeleton is a pivotal cellular structure in uni- and multi-cellular organisms: it provides architectural support and drives cargo transport, membrane deformation, and cell migration. There is evidence for age-associated defects in actin stability, dynamics, assembly, and function that impact processes including cell migration during tissue repair and regeneration, cell-cell contact during the immune response, and phagocytosis of pathogens [210, 211, 214–216].

Despite the increasing documentation of age-associated changes in the actin cytoskeleton, the underlying mechanism is not well understood. However, a possible aging mechanism targets actin filament stability. This can be seen in *C.elegans* where overexpressed calcium-binding protein, *pat-10*, increases actin stability and results in lifespan extension [218]. A critical goal of my research is to further elucidate the mechanism for controlling aging by the actin cytoskeleton using the model organism, the budding yeast, *Saccharomyces cerevisiae*. Here, we describe two projects investigating how the actin cytoskeleton declines with age and how that decline contributes to the aging phenotypes. Our findings indicate that the actin cytoskeleton is at the nexus between the biology of life and aging.

2 Age-linked decline in actin function.

In yeast, the actin cytoskeleton is organized into two main structures that persist throughout the cell cycle: actin patches and cables. Actin patches are endosomes with an F-actin coat that are essential for membrane uptake and recycling in yeast. Actin cables are bundles of F-actin that align along the mother-bud axis and extend from bud tip or bud neck to the distal tip of the mother cell. Actin cables then serve as tracks for transporting cellular constituents that are essential for bud growth and development during asymmetrical cell division in yeast.

2.1 Age-linked decline in actin stability

We obtained the first evidence that the actin cytoskeleton declines with age in the budding yeast. We isolated an aging yeast population that had completed 50% of their total replicative lifespan and used super-resolution imaging to visualize the morphology and distribution of actin cables and patches. In mother cells of advanced age, we observe a decrease in the apparent width of actin cables and an increase in the abundance of those structures (Fig. 2.1a-c).

There is evidence that declining actin filament stability may underlie the aging defects in the actin cytoskeleton seen in *C. elegans* [217, 218]. To test whether actin destabilizes with age in yeast, we challenged young and old cells with the actin destabilizing agent the Latrunculin-A (Lat-A) under conditions that preferentially destabilize actin cables without affecting actin patches or overall cell polarity. We find that actin cables undergo an age-associated decline in stability (Fig. 2.1d,e). Our findings indicate actin stability declines with age in yeast as in *C. elegans*, which raises the possibility that age-associated destabilization of actin may contribute to the aging of the actin cytoskeleton that has been documented in many organ systems in mammals and humans.

Our results also support the model that the age-associated actin cable morphological changes observed in yeast is due to destabilization of F-actin within actin cables. To test this model, we studied whether stabilization of F-actin can reverse or delay actin cable aging. Using CRISPR technology, we introduced a previously described single amino acid substitution mutation that stabilizes F-actin (*act1-159*), into the sole actin encoding *ACT1* gene of yeast [231]. We confirmed previous findings that this mutation results in an increase in actin stability. Moreover, although mutation did not extend replicative lifespan in yeast, it did have a positive effect on healthspan (Fig. 2.1g-j). Specifically, we find the stabilization of F-actin increased mitochondrial quality and promoted cell division (Fig. 2.1g-h,j). Since actin cable serves as tracks for the preferential inheritance of higher functioning mitochondria by daughter cells and transport of cellular constituents from mother cells to buds, our findings indicate that stabilization of F-actin promotes cellular healthspan and fitness by promoting actin cable function. Thus, our data suggest that controlling actin stability may be key to delay the age-associated declines in the function of the actin cytoskeleton in yeast.

2.2   Novel regulator of actin cable stability

Next, we carried out a genome-wide screen to identify genes that specifically affect actin cable stability (as opposed to stabilizing all F-actin containing structures) in yeast. Here, we screened for genes, which when deleted, reduced the sensitivity of yeast to Lat-A-induced destabilization of actin cables. A top hit revealed by our screen is an uncharacterized open reading frame, *YKL075C*. Deletion of *YKL075C* significantly suppressed Lat-A induced growth defects, potentially through effects on actin cable (Fig. 2.2c). Indeed, we found deletion of *YKL075C* resulted in stabilization of actin cables and exhibit an increase abundance of actin cables with no obvious effect upon actin cable dynamics (retrograde actin cable flow, RACF) (Fig. 2.2d-g, Supplemental Fig. 2.2c). Equally importantly, stabilization of actin cables by deletion of *YKL075C* also promoted actin cable function and cellular fitness; resulting in an increased mitochondrial function, improved cellular healthspan, and extended replicative lifespan (Fig. 2.2h-k). Thus, our study led to the discovery of a novel regulator of the actin cytoskeleton and aging in yeast.

Our studies also revealed an unexpected role for Ykl075cp in BCAA homeostasis and for BCAAs control in actin stability. Specifically, deleting *YKL075C* results in a decrease in the expression of 8 out of 10 BCAA synthetic genes and a decrease in intracellular BCAA levels (Fig. 2.3b-f). Indeed, we find that deletion of a critical BCAA biosynthetic gene (*BAT1*), or short-term depletion of a specific BCAA (leucine), results in an increase in actin cable stability, actin cable abundance, and improved mitochondrial function. Thus, we obtained the first evidence for a role for BCAA, and more specifically leucine, in control of the actin cytoskeleton and mitochondrial quality control in yeast. Finally, although the TORC1 pathway is an established sensor for amino acids including BCAAs and is a regulator of lifespan control, we find that Ykl075cp function in actin cytoskeleton control does not rely on TORC1 (Supplemental Fig. 2.3b,c). In conclusion, we identified Ykl075cp as a novel regulator of the actin cytoskeleton, BCAA metabolism, mitochondrial quality control, and aging (Fig. 5.1).

2.3   Age-linked decline in actin bundling

Our finding that actin cables undergo an age-associated decrease in apparent thickness and increase in abundance, raises the possibility that actin cable bundling may decline with age, in addition to aging effects on the stability of F-actin within actin cables. My studies indicate that the rate of RACF is similar in young and old yeast cells (Fig. 3.1b). Thus, indicating the primary drivers for RACF, F-actin polymerization and assembly into actin bundles (RACF pushing force) and myosin-powered actin cable movement (RACF pulling force), do not decline with age. Consistent with this, we find that the steady-state levels of Bni1p, one of the two formin genes of yeast that promote actin polymerization during actin cable assembly, does not change with age (Fig. 3.1a). On the other hand, we detect age-associated declines in steady-state levels of Bnr1p, a formin that mediates F-actin bundling into actin cables and actin polymerization during actin cable assembly in yeast (Fig. 3.2a,b) [4]. Since RACF rates, which are driven in part by polymerization of F-actin during actin cable assembly, are not affected, the observed decline in Bnr1p with age may contribute to lifespan control by affecting the bundling of F-actin within actin cables.

We then set forth to see if disrupting actin bundling prematurely ages a yeast cell. There are 3 actin bundling proteins in yeast (*SAC6, ABP140,* and *BNR1*). Deletion of *SAC6* (which encodes fimbrin) leads to a decrease in actin cable abundance, defects in cell polarity, and exhibit growth defects (Fig. 3.2c,d). Lastly, it also results in premature aging (Fig. 3.2e). Together, our findings support the model that actin cables unbundle with age, and that the defects in actin stability and bundling contribute to age-linked declines in actin cable function, which in turn affect yeast lifespan and healthspan (Fig. 3.3).

3   The aging mechanism of the actin cytoskeleton

In summary, we provided evidence that the actin cytoskeleton declines with age in yeast. We also propose 2 plausible theories for the mechanism(s) underlying this aspect of the aging process: destabilization and loss of bundling of F-actin within actin cables. Important goals for future studies will be to determine 1) whether actin bundling is a true aging determinant in yeast, 2) whether stabilization of F-actin within actin cables affects lifespan and healthspan, and 3) the relative contribution of actin stability and bundling to cellular fitness and lifespan in yeast.

There are many ways to reduce lifespan, which do not impact on lifespan regulating mechanisms. Indeed, since RLS is a measurement of the number of times that a mother cell can divide, any mutation or condition that inhibits cell division will result in reduced RLS. Therefore, true lifespan regulators are identified as agents, genes or conditions that extend lifespan or healthspan (cellular fitness during the aging process). To further test the theory of actin bundling as an aging determinant, we will determine whether

112

conditions that restore normal actin bundle activity in actin cables during advanced age in yeast cells extend RLS or promotes healthspan. Additionally, we will 1) determine whether Sac6p (fimbrin) protein or activity levels decline with age, 2) identify conditions in *sac6Δ* cells that compensates for any age-linked phenotypes observed, and 3) determine the effect of restoration of actin bunding activity on RLS and healthspan (MGT with age).

To determine whether stabilization of F-actin within actin cables impacts aging and quality of life during that process, we will focus on tropomyosin protein 1 (Tpm1p). Like all tropomyosins, Tpm1p binds to the lateral surface of F-actin and stabilizes actin filaments. Indeed, previous studies in our lab indicate *TPM1* overexpression results in an increase in actin cable abundance. Here too, we will 1) determine whether Tpm1p protein or activity levels decline with age, 2) identify conditions for *TPM1* expression that compensates for any age-linked phenotypes observed, and 3) determine whether promoting the stability of F-actin within actin cables also promotes RLS and healthspan.

Our findings that actin bundling declines with age raise the possibility that age-linked destabilization of the actin cytoskeleton is not just affected by the stability of actin filaments, but also by the assembly of F-actin into higher-order structures (bundling into actin cables). To determine the relative contribution of actin stability and bundling in lifespan and healthspan control in yeast, we test the effect of 1) destabilizing or stabilizing F-actin within actin cables (by regulating TPM1 expression) of an actin bundling mutant (e.g. *sac6Δ* cells) on lifespan and healthspan, and 2) promote actin bundling (by regulated *SAC6* expression) in cells with defects in the stability of F-actin within actin cables (e.g. by down-regulation of *TPM1*) on lifespan and healthpsan. While it is possible that actin stability and bundling have mutually exclusive effects on lifespan, we favor the model that these processes have complementary effects on the aging process. If this is true, 1) compromising actin bundling and stability of F-actin within actin cables will have an additive effect on actin cable abundance and function, 2) promoting actin bundling will compensate for the age-linked declines in stability of F-actin within actin cables, and 3) stabilization of F-actin within actin cables will compensate for the age-linked declines in actin bundling. Therefore, we can project a model where bundling of unstable actin filaments into actin cables is energetically unfavorable, triggering actin cables to unravel into their individual filaments and giving rise to the age-associated actin changes.

4 The functional mechanism of Ykl075cp

4.1 Ykl075cp mechanism in lifespan control

Although inhibition of BCAA biosynthesis promotes actin cable stability, it is yet to be determined whether Ykl075cp effects on lifespan are through BCAAs. Interestingly, dietary restrictions of BCAA reduce aging phenotypes and improves longevity in *Drosophila* [240]. Since our results indicate that BCAA homeostasis and *YKL075C* work in the same pathway, we then predict *bat1Δ* cells should exhibit a lifespan extension.

Similarly, although deletion of *YKL075C* promotes mitochondrial function, it is not clear whether the lifespan extension observed is due to the effects on mitochondria. Controlling mitochondrial quality is key in regulating lifespan as seen in the RACF mutants [70]. Therefore, increased mitochondrial fitness could be the underlying mechanism to extend lifespan. Testing this theory would require testing whether mtDNA loss and the associated defects in mitochondrial respiratory activity affects the lifespan extension observed in *ykl075cΔ* cells.

Finally, the mechanisms for nutrient sensing in response to the loss of *YKL075C* are not well understood. Although BCAAs are known up-stream regulators of TORC1, a well-known lifespan regulating pathway, we do not detect changes in TORC1 activity in *ykl075cΔ* cells [179–181]. Thus, Ykl075cp effects are not dependent on TORC1. It will be interesting to test whether the general amino acid control pathway (GAAC), another nutrient sensing pathway that controls lifespan, may contribute to Ykl075cp functions in controlling the actin cytoskeleton, mitochondrial quality, and lifespan. The activation of the GAAC pathway by amino acid starvation initiates the reduction of global protein synthesis and promotes translation of Gcn4p, a transcriptional activator that controls the expression of 500 genes involved in amino acid biosynthesis and aminoacyl-tRNA synthetase [138]. Gcn4p has been proven to be key in lifespan control where it is required for the extended lifespan of and globally protein repressed long-lived mutants [193, 194, 209, 288, 289]. Interestingly, the actin cytoskeleton has been implicated in regulating protein synthesis by coordinating actin bundling and GAAC through the elongation factor eEF1A [78, 79, 81, 83]. Thus, it is possible that the changes in the actin cytoskeleton observed in *ykl075cΔ* cells is through GAAC and GAAC effects on global translation and Gcn4p translation. An important goal for future studies is to determine whether 1) GAAC is activated in *ykl075cΔ* cells, 2) inhibiting GAAC blocks the effects on actin, mitochondria, and lifespan observed in *ykl075cΔ* cells, and 3) activation of GAAC promotes Ykl075cp function in the control of actin, mitochondria, and lifespan in yeast.

4.2   Underlying mechanism for BCAA homeostasis control by Ykl075cp and novel implications for the actin cytoskeleton functions

Despite our knowledge that Ykl075cp functions involve BCAA metabolism, we do not know how Ykl075cp regulates this process. Our evidence indicates Ykl075cp affects BCAA homeostasis is transcriptionally regulated through *BAT1* and *BAT2*. However, it is unlikely Ykl075cp directly functions as a transcription factor because it localizes to punctate cytoplasmic structures and does not contain conserved protein domains found in transcription factors. ChIP analysis and other DNA-binding assays are required to rule out Ykl075cp as a transcription factor. However, it is possible that Ykl075cp may act as a negative regulator of transcription. Indeed, it is possible that Ykl075cp regulates BCAA through known transcription factors for BCAA genes, like Leu3p and Gcn4p. When amino acid availability is limited, like low leucine levels, the transcriptional repressor and activator zinc-knuckle transcription factor, Leu3p, is activated through accumulated BCAA intermediates, α-isoproplymalate. This triggers the induction of *BAT1*, concurrently repressing *BAT2* expression, which are phenotypes observed upon deletion of *YKL075C* [92]. An important goal for future studies will be to identify Ykl075cp-interacting proteins and to test if those proteins affect *BAT1/2* expression and/or BCAA biosynthesis.

Changing BCAA levels directly alters actin organization. In a Maple Syrup Urine Disease (MSUD) model, toxic accumulation of BCAAs reorganizes of actin and is thought to be regulated through BCAAs interacting with RhoA [239]. We find modulating leucine levels produce phenotypes similar to those observed upon the deletion of *YKL075C*: leucine supplementation negatively alters actin's morphology; while reducing leucine levels promotes actin stability and increases actin cable abundance. These findings support a possible role for the actin cytoskeleton as a BCAA sensor and reciprocal interactions between BCAA and the actin cytoskeleton. Thus, excess BCAA levels impair actin structure and stability, which in turn affects actin function and cell fitness. Conversely, amino acid sensing pathways may then modulate BCAA homeostasis to promote actin cytoskeleton organization and function when amino acids are limiting, consequently promoting mitochondrial fitness and primes the cell for improved longevity.

5   Figures

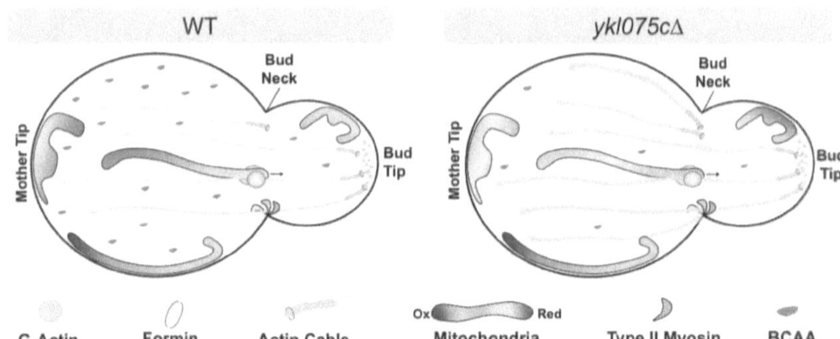

Figure 5.1: Ykl075cp regulates actin through changing BCAA metabolism to control lifespan.
Ykl075cp regulates the actin cytoskeleton and mitochondria quality to improve lifespan expectancy in yeast. Alters BCAA metabolism to reduce BCAA levels, which subsequently induces increased actin cable stability and abundance, to enhance mitochondrial quality and extends yeast lifespan.

References

1. Davidson, A. J., and Wood, W. (2016). Unravelling the actin cytoskeleton: A new competitive edge? Trends Cell Biol. *26*, 569–576.

2. Lai, W.-F., and Wong, W.-T. (2020). Roles of the actin cytoskeleton in aging and age-associated diseases. Ageing Res Rev *58*, 101021.

3. Svitkina, T. (2018). The Actin Cytoskeleton and Actin-Based Motility. Cold Spring Harb. Perspect. Biol. *10*.

4. Moseley, J. B., and Goode, B. L. (2006). The yeast actin cytoskeleton: from cellular function to biochemical mechanism. Microbiol. Mol. Biol. Rev. *70*, 605–645.

5. Kabsch, W., Mannherz, H. G., Suck, D., Pai, E. F., and Holmes, K. C. (1990). Atomic structure of the actin:DNase I complex. Nature *347*, 37–44.

6. Dominguez, R., and Holmes, K. C. (2011). Actin structure and function. Annu. Rev. Biophys. *40*, 169–186.

7. De La Cruz, E. M., Mandinova, A., Steinmetz, M. O., Stoffler, D., Aebi, U., and Pollard, T. D. (2000). Polymerization and structure of nucleotide-free actin filaments. J. Mol. Biol. *295*, 517–526.

8. Carlier, M. F., Valentin-Ranc, C., Combeau, C., Fievez, S., and Pantoloni, D. (1994). Actin polymerization: regulation by divalent metal ion and nucleotide binding, ATP hydrolysis and binding of myosin. Adv. Exp. Med. Biol. *358*, 71–81.

9. Pollard, T. D., and Borisy, G. G. (2003). Cellular motility driven by assembly and disassembly of actin filaments. Cell *112*, 453–465.

10. Small, J. V., Isenberg, G., and Celis, J. E. (1978). Polarity of actin at the leading edge of cultured cells. Nature *272*, 638–639.

11. Pollard, T. D., Blanchoin, L., and Mullins, R. D. (2000). Molecular mechanisms controlling actin filament dynamics in nonmuscle cells. Annu. Rev. Biophys. Biomol. Struct. *29*, 545–576.

12. Carlier, M. F., and Pantaloni, D. (1986). Direct evidence for ADP-Pi-F-actin as the major intermediate in ATP-actin polymerization. Rate of dissociation of Pi from actin filaments. Biochemistry *25*, 7789–7792.

13. Blanchoin, L., and Pollard, T. D. (2002). Hydrolysis of ATP by polymerized actin depends on the bound divalent cation but not profilin. Biochemistry *41*, 597–602.

14. Pollard, T. D. (2016). Actin and Actin-Binding Proteins. Cold Spring Harb. Perspect. Biol. *8*.

15. Robinson, R. C., Turbedsky, K., Kaiser, D. A., Marchand, J. B., Higgs, H. N., Choe, S., and Pollard, T. D. (2001). Crystal structure of Arp2/3 complex. Science *294*, 1679–1684.

16. Rottner, K., Hänisch, J., and Campellone, K. G. (2010). WASH, WHAMM and JMY: regulation of Arp2/3 complex and beyond. Trends Cell Biol. *20*, 650–661.

17. Rotty, J. D., Wu, C., and Bear, J. E. (2013). New insights into the regulation and cellular functions of the ARP2/3 complex. Nat. Rev. Mol. Cell Biol. *14*, 7–12.

18. Rouiller, I., Xu, X.-P., Amann, K. J., Egile, C., Nickell, S., Nicastro, D., Li, R., Pollard, T. D., Volkmann, N., and Hanein, D. (2008). The structural basis of actin filament branching by the Arp2/3 complex. J. Cell Biol. *180*, 887–895.

19. Ti, S.-C., Jurgenson, C. T., Nolen, B. J., and Pollard, T. D. (2011). Structural and biochemical characterization of two binding sites for nucleation-promoting factor WASp-VCA on Arp2/3 complex. Proc. Natl. Acad. Sci. USA *108*, E463-71.

20. Huckaba, T. M., Gay, A. C., Pantalena, L. F., Yang, H.-C., and Pon, L. A. (2004). Live cell imaging of the assembly, disassembly, and actin cable-dependent movement of endosomes and actin patches in the budding yeast, Saccharomyces cerevisiae. J. Cell Biol. *167*, 519–530.

21. Evangelista, M., Zigmond, S., and Boone, C. (2003). Formins: signaling effectors for assembly and polarization of actin filaments. J. Cell Sci. *116*, 2603–2611.

22. Wallar, B. J., and Alberts, A. S. (2003). The formins: active scaffolds that remodel the cytoskeleton. Trends Cell Biol. *13*, 435–446.

23. Alberts, A. S. (2001). Identification of a carboxyl-terminal diaphanous-related formin homology protein autoregulatory domain. J. Biol. Chem. *276*, 2824–2830.

24. Otomo, T., Tomchick, D. R., Otomo, C., Machius, M., and Rosen, M. K. (2010). Crystal structure of the Formin mDia1 in autoinhibited conformation. PLoS One *5*.

25. Nezami, A., Poy, F., Toms, A., Zheng, W., and Eck, M. J. (2010). Crystal structure of a complex between amino and carboxy terminal fragments of mDia1: insights into autoinhibition of diaphanous-related formins. PLoS One *5*.

26. Breitsprecher, D., and Goode, B. L. (2013). Formins at a glance. J. Cell Sci. *126*, 1–7.

27. Watanabe, N., Kato, T., Fujita, A., Ishizaki, T., and Narumiya, S. (1999). Cooperation between mDia1 and ROCK in Rho-induced actin reorganization. Nat. Cell Biol. *1*, 136–143.

28. Goode, B. L., and Eck, M. J. (2007). Mechanism and function of formins in the control of actin assembly. Annu. Rev. Biochem. *76*, 593–627.

29. Otomo, T., Tomchick, D. R., Otomo, C., Panchal, S. C., Machius, M., and Rosen, M. K. (2005). Structural basis of actin filament nucleation and processive capping by a formin homology 2 domain. Nature *433*, 488–494.

30. Xu, Y., Moseley, J. B., Sagot, I., Poy, F., Pellman, D., Goode, B. L., and Eck, M. J. (2004). Crystal structures of a Formin Homology-2 domain reveal a tethered dimer architecture. Cell *116*, 711–723.

31. Kozlov, M. M., and Bershadsky, A. D. (2004). Processive capping by formin suggests a force-driven mechanism of actin polymerization. J. Cell Biol. *167*, 1011–1017.

32. Moseley, J. B., Sagot, I., Manning, A. L., Xu, Y., Eck, M. J., Pellman, D., and Goode, B. L. (2004). A conserved mechanism for Bni1- and mDia1-induced actin assembly and dual regulation of Bni1 by Bud6 and profilin. Mol. Biol. Cell *15*, 896–907.

33. Romero, S., Le Clainche, C., Didry, D., Egile, C., Pantaloni, D., and Carlier, M.-F. (2004). Formin is a processive motor that requires profilin to accelerate actin assembly and associated ATP hydrolysis. Cell *119*, 419–429.

34. Kovar, D. R., Harris, E. S., Mahaffy, R., Higgs, H. N., and Pollard, T. D. (2006). Control of the assembly of ATP- and ADP-actin by formins and profilin. Cell *124*, 423–435.

35. Vavylonis, D., Kovar, D. R., O'Shaughnessy, B., and Pollard, T. D. (2006). Model of formin-associated actin filament elongation. Mol. Cell *21*, 455–466.

36. Paul, A. S., and Pollard, T. D. (2009). Review of the mechanism of processive actin filament elongation by formins. Cell Motil. Cytoskeleton *66*, 606–617.

37. Wolven, A. K., Belmont, L. D., Mahoney, N. M., Almo, S. C., and Drubin, D. G. (2000). In vivo importance of actin nucleotide exchange catalyzed by profilin. J. Cell Biol. *150*, 895–904.

38. Pring, M., Evangelista, M., Boone, C., Yang, C., and Zigmond, S. H. (2003). Mechanism of formin-induced nucleation of actin filaments. Biochemistry *42*, 486–496.

39. Neidt, E. M., Scott, B. J., and Kovar, D. R. (2009). Formin differentially utilizes profilin isoforms to rapidly assemble actin filaments. J. Biol. Chem. *284*, 673–684.

40. Courtemanche, N., and Pollard, T. D. (2012). Determinants of Formin Homology 1 (FH1) domain function in actin filament elongation by formins. J. Biol. Chem. *287*, 7812–7820.

41. Shortle, D., Haber, J. E., and Botstein, D. (1982). Lethal disruption of the yeast actin gene by integrative DNA transformation. Science *217*, 371–373.

42. Engqvist-Goldstein, A. E. Y., and Drubin, D. G. (2003). Actin assembly and endocytosis: from yeast to mammals. Annu. Rev. Cell Dev. Biol. *19*, 287–332.

43. Spiering, D., and Hodgson, L. (2011). Dynamics of the Rho-family small GTPases in actin regulation and motility. Cell Adh Migr *5*, 170–180.

44. Ozaki-Kuroda, K., Yamamoto, Y., Nohara, H., Kinoshita, M., Fujiwara, T., Irie, K., and Takai, Y. (2001). Dynamic localization and function of Bni1p at the sites of directed growth in Saccharomyces cerevisiae. Mol. Cell. Biol. *21*, 827–839.

45. Imamura, H., Tanaka, K., Hihara, T., Umikawa, M., Kamei, T., Takahashi, K., Sasaki, T., and Takai, Y. (1997). Bni1p and Bnr1p: downstream targets of the Rho family small G-proteins which interact with profilin and regulate actin cytoskeleton in Saccharomyces cerevisiae. EMBO J. *16*, 2745–2755.

46. Evangelista, M., Blundell, K., Longtine, M. S., Chow, C. J., Adames, N., Pringle, J. R., Peter, M., and Boone, C. (1997). Bni1p, a yeast formin linking cdc42p and the actin cytoskeleton during polarized morphogenesis. Science *276*, 118–122.

47. Cooper, J. A. (2002). Actin dynamics: tropomyosin provides stability. Curr. Biol. *12*, R523-5.

48. Cao, W., Goodarzi, J. P., and De La Cruz, E. M. (2006). Energetics and kinetics of cooperative cofilin-actin filament interactions. J. Mol. Biol. *361*, 257–267.

49. Hayden, S. M., Miller, P. S., Brauweiler, A., and Bamburg, J. R. (1993). Analysis of the interactions of actin depolymerizing factor with G- and F-actin. Biochemistry *32*, 9994–10004.

50. Bobkov, A. A., Muhlrad, A., Shvetsov, A., Benchaar, S., Scoville, D., Almo, S. C., and Reisler, E. (2004). Cofilin (ADF) affects lateral contacts in F-actin. J. Mol. Biol. *337*, 93–104.

51. Galkin, V. E., Orlova, A., Kudryashov, D. S., Solodukhin, A., Reisler, E., Schröder, G. F., and Egelman, E. H. (2011). Remodeling of actin filaments by ADF/cofilin proteins. Proc. Natl. Acad. Sci. USA *108*, 20568–20572.

52. McCullough, B. R., Grintsevich, E. E., Chen, C. K., Kang, H., Hutchison, A. L., Henn, A., Cao, W., Suarez, C., Martiel, J.-L., Blanchoin, L., et al. (2011). Cofilin-linked changes in actin filament flexibility

promote severing. Biophys. J. *101*, 151–159.

53. Okada, K., Ravi, H., Smith, E. M., and Goode, B. L. (2006). Aip1 and cofilin promote rapid turnover of yeast actin patches and cables: a coordinated mechanism for severing and capping filaments. Mol. Biol. Cell *17*, 2855–2868.

54. Schott, D. H., Collins, R. N., and Bretscher, A. (2002). Secretory vesicle transport velocity in living cells depends on the myosin-V lever arm length. J. Cell Biol. *156*, 35–39.

55. Darzacq, X., Powrie, E., Gu, W., Singer, R. H., and Zenklusen, D. (2003). RNA asymmetric distribution and daughter/mother differentiation in yeast. Curr. Opin. Microbiol. *6*, 614–620.

56. Fehrenbacher, K. L., Yang, H.-C., Gay, A. C., Huckaba, T. M., and Pon, L. A. (2004). Live cell imaging of mitochondrial movement along actin cables in budding yeast. Curr. Biol. *14*, 1996–2004.

57. Yang, H.-C., and Pon, L. A. (2002). Actin cable dynamics in budding yeast. Proc. Natl. Acad. Sci. USA *99*, 751–756.

58. Huckaba, T. M., Lipkin, T., and Pon, L. A. (2006). Roles of type II myosin and a tropomyosin isoform in retrograde actin flow in budding yeast. J. Cell Biol. *175*, 957–969.

59. Lidke, D. S., Lidke, K. A., Rieger, B., Jovin, T. M., and Arndt-Jovin, D. J. (2005). Reaching out for signals: filopodia sense EGF and respond by directed retrograde transport of activated receptors. J. Cell Biol. *170*, 619–626.

60. Zhu, R., Liu, C., and Gundersen, G. G. (2018). Nuclear positioning in migrating fibroblasts. Semin. Cell Dev. Biol. *82*, 41–50.

61. Brown, M. E., and Bridgman, P. C. (2003). Retrograde flow rate is increased in growth cones from myosin IIB knockout mice. J. Cell Sci. *116*, 1087–1094.

62. Anderson, T. W., Vaughan, A. N., and Cramer, L. P. (2008). Retrograde flow and myosin II activity within the leading cell edge deliver F-actin to the lamella to seed the formation of graded polarity actomyosin II filament bundles in migrating fibroblasts. Mol. Biol. Cell *19*, 5006–5018.

63. Abercrombie, M., Heaysman, J. E., and Pegrum, S. M. (1970). The locomotion of fibroblasts in culture. I. Movements of the leading edge. Exp. Cell Res. *59*, 393–398.

64. Bornschlögl, T., Romero, S., Vestergaard, C. L., Joanny, J.-F., Van Nhieu, G. T., and Bassereau, P. (2013). Filopodial retraction force is generated by cortical actin dynamics and controlled by reversible tethering at the tip. Proc. Natl. Acad. Sci. USA *110*, 18928–18933.

65. Hu, K., Ji, L., Applegate, K. T., Danuser, G., and Waterman-Storer, C. M. (2007). Differential transmission of actin motion within focal adhesions. Science *315*, 111–115.

66. Sherer, N. M., Lehmann, M. J., Jimenez-Soto, L. F., Horensavitz, C., Pypaert, M., and Mothes, W. (2007). Retroviruses can establish filopodial bridges for efficient cell-to-cell transmission. Nat. Cell Biol. *9*, 310–315.

67. Kaplan, G. (1977). Differences in the mode of phagocytosis with Fc and C3 receptors in macrophages. Scand. J. Immunol. *6*, 797–807.

68. Rougerie, P., Miskolci, V., and Cox, D. (2013). Generation of membrane structures during phagocytosis and chemotaxis of macrophages: role and regulation of the actin cytoskeleton. Immunol. Rev. *256*, 222–239.

69. Araki, N., Hatae, T., Furukawa, A., and Swanson, J. A. (2003). Phosphoinositide-3-kinase-independent contractile activities associated with Fcgamma-receptor-mediated phagocytosis and macropinocytosis in macrophages. J. Cell Sci. *116*, 247–257.

70. Higuchi, R., Vevea, J. D., Swayne, T. C., Chojnowski, R., Hill, V., Boldogh, I. R., and Pon, L. A. (2013). Actin dynamics affect mitochondrial quality control and aging in budding yeast. Curr. Biol. *23*, 2417–2422.

71. McFaline-Figueroa, J. R., Vevea, J., Swayne, T. C., Zhou, C., Liu, C., Leung, G., Boldogh, I. R., and Pon, L. A. (2011). Mitochondrial quality control during inheritance is associated with lifespan and mother-daughter age asymmetry in budding yeast. Aging Cell *10*, 885–895.

72. Katajisto, P., Döhla, J., Chaffer, C. L., Pentinmikko, N., Marjanovic, N., Iqbal, S., Zoncu, R., Chen, W., Weinberg, R. A., and Sabatini, D. M. (2015). Stem cells. Asymmetric apportioning of aged mitochondria between daughter cells is required for stemness. Science *348*, 340–343.

73. Pavitt, G. D., Ramaiah, K. V., Kimball, S. R., and Hinnebusch, A. G. (1998). eIF2 independently binds two distinct eIF2B subcomplexes that catalyze and regulate guanine-nucleotide exchange. Genes Dev. *12*, 514–526.

74. Dever, T. E., Feng, L., Wek, R. C., Cigan, A. M., Donahue, T. F., and Hinnebusch, A. G. (1992). Phosphorylation of initiation factor 2 alpha by protein kinase GCN2 mediates gene-specific translational control of GCN4 in yeast. Cell *68*, 585–596.

75. Mateyak, M. K., and Kinzy, T. G. (2010). eEF1A: thinking outside the ribosome. J. Biol. Chem. *285*, 21209–21213.

76. Yang, F., Demma, M., Warren, V., Dharmawardhane, S., and Condeelis, J. (1990). Identification of an actin-binding protein from Dictyostelium as elongation factor 1a. Nature *347*, 494–496.

77. Munshi, R., Kandl, K. A., Carr-Schmid, A., Whitacre, J. L., Adams, A. E., and Kinzy, T. G. (2001). Overexpression of translation elongation factor 1A affects the organization and function of the actin cytoskeleton in yeast. Genetics *157*, 1425–1436.

78. Owen, C. H., DeRosier, D. J., and Condeelis, J. (1992). Actin crosslinking protein EF-1a of Dictyostelium discoideum has a unique bonding rule that allows square-packed bundles. J. Struct. Biol. *109*, 248–254.

79. Gross, S. R., and Kinzy, T. G. (2007). Improper organization of the actin cytoskeleton affects protein synthesis at initiation. Mol. Cell. Biol. *27*, 1974–1989.

80. Gross, S. R., and Kinzy, T. G. (2005). Translation elongation factor 1A is essential for regulation of the actin cytoskeleton and cell morphology. Nat. Struct. Mol. Biol. *12*, 772–778.

81. Perez, W. B., and Kinzy, T. G. (2014). Translation elongation factor 1A mutants with altered actin bundling activity show reduced aminoacyl-tRNA binding and alter initiation via eIF2α phosphorylation. J. Biol. Chem. *289*, 20928–20938.

82. Edmonds, B. T., Bell, A., Wyckoff, J., Condeelis, J., and Leyh, T. S. (1998). The effect of F-actin on the binding and hydrolysis of guanine nucleotide by Dictyostelium elongation factor 1A. J. Biol. Chem. *273*, 10288–10295.

83. Visweswaraiah, J., Lageix, S., Castilho, B. A., Izotova, L., Kinzy, T. G., Hinnebusch, A. G., and Sattlegger, E. (2011). Evidence that eukaryotic translation elongation factor 1A (eEF1A) binds the Gcn2 protein C terminus and inhibits Gcn2 activity. J. Biol. Chem. *286*, 36568–36579.

84. Sattlegger, E., Swanson, M. J., Ashcraft, E. A., Jennings, J. L., Fekete, R. A., Link, A. J., and Hinnebusch, A. G. (2004). YIH1 is an actin-binding protein that inhibits protein kinase GCN2 and impairs general amino acid control when overexpressed. J. Biol. Chem. *279*, 29952–29962.

85. Sattlegger, E., Barbosa, J. A. R. G., Moraes, M. C. S., Martins, R. M., Hinnebusch, A. G., and Castilho, B. A. (2011). Gcn1 and actin binding to Yih1: implications for activation of the eIF2 kinase GCN2. J. Biol. Chem. *286*, 10341–10355.

86. Pereira, C. M., Sattlegger, E., Jiang, H.-Y., Longo, B. M., Jaqueta, C. B., Hinnebusch, A. G., Wek, R. C., Mello, L. E. A. M., and Castilho, B. A. (2005). IMPACT, a protein preferentially expressed in the mouse brain, binds GCN1 and inhibits GCN2 activation. J. Biol. Chem. *280*, 28316–28323.

87. Cambiaghi, T. D., Pereira, C. M., Shanmugam, R., Bolech, M., Wek, R. C., Sattlegger, E., and Castilho, B. A. (2014). Evolutionarily conserved IMPACT impairs various stress responses that require GCN1 for activating the eIF2 kinase GCN2. Biochem. Biophys. Res. Commun. *443*, 592–597.

88. Silva, R. C., Sattlegger, E., and Castilho, B. A. (2016). Perturbations in actin dynamics reconfigure protein complexes that modulate GCN2 activity and promote an eIF2 response. J. Cell Sci. *129*, 4521–4533.

89. Yudkoff, M. (1997). Brain metabolism of branched-chain amino acids. Glia *21*, 92–98.

90. Lynch, C. J., and Adams, S. H. (2014). Branched-chain amino acids in metabolic signalling and insulin resistance. Nat. Rev. Endocrinol. *10*, 723–736.

91. Lake, A. D., Novak, P., Shipkova, P., Aranibar, N., Robertson, D. G., Reily, M. D., Lehman-McKeeman, L. D., Vaillancourt, R. R., and Cherrington, N. J. (2015). Branched chain amino acid metabolism profiles in progressive human nonalcoholic fatty liver disease. Amino Acids *47*, 603–615.

92. Kohlhaw, G. B. (2003). Leucine biosynthesis in fungi: entering metabolism through the back door. Microbiol. Mol. Biol. Rev. *67*, 1–15, table of contents.

93. Jastrzębowska, K., and Gabriel, I. (2015). Inhibitors of amino acids biosynthesis as antifungal agents. Amino Acids *47*, 227–249.

94. López, G., Quezada, H., Duhne, M., González, J., Lezama, M., El-Hafidi, M., Colón, M., Martínez de la Escalera, X., Flores-Villegas, M. C., Scazzocchio, C., et al. (2015). Diversification of Paralogous α-Isopropylmalate Synthases by Modulation of Feedback Control and Hetero-Oligomerization in Saccharomyces cerevisiae. Eukaryotic Cell *14*, 564–577.

95. Mewes, H. W., Albermann, K., Bähr, M., Frishman, D., Gleissner, A., Hani, J., Heumann, K., Kleine, K., Maierl, A., Oliver, S. G., et al. (1997). Overview of the yeast genome. Nature *387*, 7–65.

96. Colón, M., Hernández, F., López, K., Quezada, H., González, J., López, G., Aranda, C., and González, A. (2011). Saccharomyces cerevisiae Bat1 and Bat2 aminotransferases have functionally diverged from the ancestral-like Kluyveromyces lactis orthologous enzyme. PLoS One *6*, e16099.

97. González, J., López, G., Argueta, S., Escalera-Fanjul, X., El Hafidi, M., Campero-Basaldua, C., Strauss, J., Riego-Ruiz, L., and González, A. (2017). Diversification of Transcriptional Regulation Determines Subfunctionalization of Paralogous Branched Chain Aminotransferases in the Yeast Saccharomyces cerevisiae. Genetics *207*, 975–991.

98. Takpho, N., Watanabe, D., and Takagi, H. (2018). Valine biosynthesis in Saccharomyces cerevisiae is regulated by the mitochondrial branched-chain amino acid aminotransferase Bat1. Microb. Cell *5*, 293–299.

99. Kira, S., Kumano, Y., Ukai, H., Takeda, E., Matsuura, A., and Noda, T. (2016). Dynamic relocation of the TORC1-Gtr1/2-Ego1/2/3 complex is regulated by Gtr1 and Gtr2. Mol. Biol. Cell *27*, 382–396.

100. Wullschleger, S., Loewith, R., and Hall, M. N. (2006). TOR signaling in growth and metabolism. Cell *124*, 471–484.

101. Loewith, R., and Hall, M. N. (2011). Target of rapamycin (TOR) in nutrient signaling and growth control. Genetics *189*, 1177–1201.

102. Heitman, J., Movva, N. R., and Hall, M. N. (1991). Targets for cell cycle arrest by the immunosuppressant rapamycin in yeast. Science *253*, 905–909.

103. Kunz, J., Henriquez, R., Schneider, U., Deuter-Reinhard, M., Movva, N. R., and Hall, M. N. (1993). Target of rapamycin in yeast, TOR2, is an essential phosphatidylinositol kinase homolog required for G1 progression. Cell *73*, 585–596.

104. Schmelzle, T., and Hall, M. N. (2000). TOR, a central controller of cell growth. Cell *103*, 253–262.

105. Schreiber, S. L. (1991). Chemistry and biology of the immunophilins and their immunosuppressive ligands. Science *251*, 283–287.

106. Conrad, M., Schothorst, J., Kankipati, H. N., Van Zeebroeck, G., Rubio-Texeira, M., and Thevelein, J. M. (2014). Nutrient sensing and signaling in the yeast Saccharomyces cerevisiae. FEMS Microbiol. Rev. *38*, 254–299.

107. Loewith, R., Jacinto, E., Wullschleger, S., Lorberg, A., Crespo, J. L., Bonenfant, D., Oppliger, W., Jenoe, P., and Hall, M. N. (2002). Two TOR complexes, only one of which is rapamycin sensitive, have distinct roles in cell growth control. Mol. Cell *10*, 457–468.

108. Wedaman, K. P., Reinke, A., Anderson, S., Yates, J., McCaffery, J. M., and Powers, T. (2003). Tor kinases are in distinct membrane-associated protein complexes in Saccharomyces cerevisiae. Mol. Biol. Cell *14*, 1204–1220.

109. Reinke, A., Anderson, S., McCaffery, J. M., Yates, J., Aronova, S., Chu, S., Fairclough, S., Iverson, C., Wedaman, K. P., and Powers, T. (2004). TOR complex 1 includes a novel component, Tco89p (YPL180w), and cooperates with Ssd1p to maintain cellular integrity in Saccharomyces cerevisiae. J. Biol. Chem. *279*, 14752–14762.

110. Chen, E. J., and Kaiser, C. A. (2003). LST8 negatively regulates amino acid biosynthesis as a component of the TOR pathway. J. Cell Biol. *161*, 333–347.

111. Kim, D.-H., Sarbassov, D. D., Ali, S. M., Latek, R. R., Guntur, K. V. P., Erdjument-Bromage, H., Tempst, P., and Sabatini, D. M. (2003). GbetaL, a positive regulator of the rapamycin-sensitive pathway required for the nutrient-sensitive interaction between raptor and mTOR. Mol. Cell *11*, 895–904.

112. Binda, M., Péli-Gulli, M.-P., Bonfils, G., Panchaud, N., Urban, J., Sturgill, T. W., Loewith, R., and De Virgilio, C. (2009). The Vam6 GEF controls TORC1 by activating the EGO complex. Mol. Cell *35*, 563–573.

113. Hatakeyama, R., Péli-Gulli, M.-P., Hu, Z., Jaquenoud, M., Garcia Osuna, G. M., Sardu, A., Dengjel, J., and De Virgilio, C. (2019). Spatially distinct pools of TORC1 balance protein homeostasis. Mol. Cell *73*, 325–338.e8.

114. Urban, J., Soulard, A., Huber, A., Lippman, S., Mukhopadhyay, D., Deloche, O., Wanke, V., Anrather,

D., Ammerer, G., Riezman, H., et al. (2007). Sch9 is a major target of TORC1 in Saccharomyces cerevisiae. Mol. Cell *26*, 663–674.

115. Roth, A. F., Wan, J., Bailey, A. O., Sun, B., Kuchar, J. A., Green, W. N., Phinney, B. S., Yates, J. R., and Davis, N. G. (2006). Global analysis of protein palmitoylation in yeast. Cell *125*, 1003–1013.

116. Nadolski, M. J., and Linder, M. E. (2009). Molecular recognition of the palmitoylation substrate Vac8 by its palmitoyltransferase Pfa3. J. Biol. Chem. *284*, 17720–17730.

117. Powis, K., Zhang, T., Panchaud, N., Wang, R., De Virgilio, C., and Ding, J. (2015). Crystal structure of the Ego1-Ego2-Ego3 complex and its role in promoting Rag GTPase-dependent TORC1 signaling. Cell Res. *25*, 1043–1059.

118. Gao, M., and Kaiser, C. A. (2006). A conserved GTPase-containing complex is required for intracellular sorting of the general amino-acid permease in yeast. Nat. Cell Biol. *8*, 657–667.

119. Neklesa, T. K., and Davis, R. W. (2009). A genome-wide screen for regulators of TORC1 in response to amino acid starvation reveals a conserved Npr2/3 complex. PLoS Genet. *5*, e1000515.

120. Panchaud, N., Péli-Gulli, M.-P., and De Virgilio, C. (2013). SEACing the GAP that nEGOCiates TORC1 activation: evolutionary conservation of Rag GTPase regulation. Cell Cycle *12*, 2948–2952.

121. Hughes Hallett, J. E., Luo, X., and Capaldi, A. P. (2014). State transitions in the TORC1 signaling pathway and information processing in Saccharomyces cerevisiae. Genetics *198*, 773–786.

122. Dokudovskaya, S., and Rout, M. P. (2015). SEA you later alli-GATOR--a dynamic regulator of the TORC1 stress response pathway. J. Cell Sci. *128*, 2219–2228.

123. Péli-Gulli, M.-P., Sardu, A., Panchaud, N., Raucci, S., and De Virgilio, C. (2015). Amino Acids Stimulate TORC1 through Lst4-Lst7, a GTPase-Activating Protein Complex for the Rag Family GTPase Gtr2. Cell Rep. *13*, 1–7.

124. Broach, J. R. (2012). Nutritional control of growth and development in yeast. Genetics *192*, 73–105.

125. Wanke, V., Cameroni, E., Uotila, A., Piccolis, M., Urban, J., Loewith, R., and De Virgilio, C. (2008). Caffeine extends yeast lifespan by targeting TORC1. Mol. Microbiol. *69*, 277–285.

126. Huber, A., French, S. L., Tekotte, H., Yerlikaya, S., Stahl, M., Perepelkina, M. P., Tyers, M., Rougemont, J., Beyer, A. L., and Loewith, R. (2011). Sch9 regulates ribosome biogenesis via Stb3, Dot6 and Tod6 and the histone deacetylase complex RPD3L. EMBO J. *30*, 3052–3064.

127. Lempiäinen, H., Uotila, A., Urban, J., Dohnal, I., Ammerer, G., Loewith, R., and Shore, D. (2009). Sfp1 interaction with TORC1 and Mrs6 reveals feedback regulation on TOR signaling. Mol. Cell *33*, 704–716.

128. González, A., Shimobayashi, M., Eisenberg, T., Merle, D. A., Pendl, T., Hall, M. N., and Moustafa, T. (2015). TORC1 promotes phosphorylation of ribosomal protein S6 via the AGC kinase Ypk3 in Saccharomyces cerevisiae. PLoS One *10*, e0120250.

129. Düvel, K., Santhanam, A., Garrett, S., Schneper, L., and Broach, J. R. (2003). Multiple roles of Tap42 in mediating rapamycin-induced transcriptional changes in yeast. Mol. Cell *11*, 1467–1478.

130. Zheng, Y., and Jiang, Y. (2005). The yeast phosphotyrosyl phosphatase activator is part of the Tap42-phosphatase complexes. Mol. Biol. Cell *16*, 2119–2127.

131. Shamji, A. F., Kuruvilla, F. G., and Schreiber, S. L. (2000). Partitioning the transcriptional program

induced by rapamycin among the effectors of the Tor proteins. Curr. Biol. *10*, 1574–1581.

132. Loewith, R. (2010). TORC1 signaling in budding yeast. In The Enzymes. (Elsevier), pp. 147–175.

133. Bonfils, G., Jaquenoud, M., Bontron, S., Ostrowicz, C., Ungermann, C., and De Virgilio, C. (2012). Leucyl-tRNA synthetase controls TORC1 via the EGO complex. Mol. Cell *46*, 105–110.

134. González, A., and Hall, M. N. (2017). Nutrient sensing and TOR signaling in yeast and mammals. EMBO J. *36*, 397–408.

135. Kingsbury, J. M., Sen, N. D., and Cardenas, M. E. (2015). Branched-Chain Aminotransferases Control TORC1 Signaling in Saccharomyces cerevisiae. PLoS Genet. *11*, e1005714.

136. Nicklin, P., Bergman, P., Zhang, B., Triantafellow, E., Wang, H., Nyfeler, B., Yang, H., Hild, M., Kung, C., Wilson, C., et al. (2009). Bidirectional transport of amino acids regulates mTOR and autophagy. Cell *136*, 521–534.

137. Wolfson, R. L., Chantranupong, L., Saxton, R. A., Shen, K., Scaria, S. M., Cantor, J. R., and Sabatini, D. M. (2016). Sestrin2 is a leucine sensor for the mTORC1 pathway. Science *351*, 43–48.

138. Hinnebusch, A. G. (2005). Translational regulation of GCN4 and the general amino acid control of yeast. Annu. Rev. Microbiol. *59*, 407–450.

139. Akram, Z., Ahmed, I., Mack, H., Kaur, R., Silva, R. C., Castilho, B. A., Friant, S., Sattlegger, E., and Munn, A. L. (2020). Yeast as a Model to Understand Actin-Mediated Cellular Functions in Mammals-Illustrated with Four Actin Cytoskeleton Proteins. Cells *9*.

140. Romano, P. R., Garcia-Barrio, M. T., Zhang, X., Wang, Q., Taylor, D. R., Zhang, F., Herring, C., Mathews, M. B., Qin, J., and Hinnebusch, A. G. (1998). Autophosphorylation in the activation loop is required for full kinase activity in vivo of human and yeast eukaryotic initiation factor 2alpha kinases PKR and GCN2. Mol. Cell. Biol. *18*, 2282–2297.

141. Qiu, H., Dong, J., Hu, C., Francklyn, C. S., and Hinnebusch, A. G. (2001). The tRNA-binding moiety in GCN2 contains a dimerization domain that interacts with the kinase domain and is required for tRNA binding and kinase activation. EMBO J. *20*, 1425–1438.

142. Hinnebusch, A. G. (2005). eIF2alpha kinases provide a new solution to the puzzle of substrate specificity. Nat. Struct. Mol. Biol. *12*, 835–838.

143. Gárriz, A., Qiu, H., Dey, M., Seo, E.-J., Dever, T. E., and Hinnebusch, A. G. (2009). A network of hydrophobic residues impeding helix alphaC rotation maintains latency of kinase Gcn2, which phosphorylates the alpha subunit of translation initiation factor 2. Mol. Cell. Biol. *29*, 1592–1607.

144. Lageix, S., Rothenburg, S., Dever, T. E., and Hinnebusch, A. G. (2014). Enhanced interaction between pseudokinase and kinase domains in Gcn2 stimulates eIF2α phosphorylation in starved cells. PLoS Genet. *10*, e1004326.

145. Lageix, S., Zhang, J., Rothenburg, S., and Hinnebusch, A. G. (2015). Interaction between the tRNA-binding and C-terminal domains of Yeast Gcn2 regulates kinase activity in vivo. PLoS Genet. *11*, e1004991.

146. Marton, M. J., Vazquez de Aldana, C. R., Qiu, H., Chakraburtty, K., and Hinnebusch, A. G. (1997). Evidence that GCN1 and GCN20, translational regulators of GCN4, function on elongating ribosomes in activation of eIF2alpha kinase GCN2. Mol. Cell. Biol. *17*, 4474–4489.

147. Ramirez, M., Wek, R. C., and Hinnebusch, A. G. (1991). Ribosome association of GCN2 protein

kinase, a translational activator of the GCN4 gene of Saccharomyces cerevisiae. Mol. Cell. Biol. *11*, 3027–3036.

148. Sattlegger, E., and Hinnebusch, A. G. (2000). Separate domains in GCN1 for binding protein kinase GCN2 and ribosomes are required for GCN2 activation in amino acid-starved cells. EMBO J. *19*, 6622–6633.

149. Müller, I., Zimmermann, M., Becker, D., and Flömer, M. (1980). Calendar life span versus budding lifespan of Saccharomyces cerevisiae. Mech. Ageing Dev. *12*, 47–52.

150. Mortimer, R. K., and Johnston, J. R. (1959). Life span of individual yeast cells. Nature *183*, 1751–1752.

151. Hughes, A. L., and Gottschling, D. E. (2012). An early age increase in vacuolar pH limits mitochondrial function and lifespan in yeast. Nature *492*, 261–265.

152. Howitz, K. T., Bitterman, K. J., Cohen, H. Y., Lamming, D. W., Lavu, S., Wood, J. G., Zipkin, R. E., Chung, P., Kisielewski, A., Zhang, L.-L., et al. (2003). Small molecule activators of sirtuins extend Saccharomyces cerevisiae lifespan. Nature *425*, 191–196.

153. Erjavec, N., Cvijovic, M., Klipp, E., and Nyström, T. (2008). Selective benefits of damage partitioning in unicellular systems and its effects on aging. Proc. Natl. Acad. Sci. USA *105*, 18764–18769.

154. Steffen, K. K., Kennedy, B. K., and Kaeberlein, M. (2009). Measuring replicative life span in the budding yeast. J. Vis. Exp.

155. Chen, K. L., Crane, M. M., and Kaeberlein, M. (2017). Microfluidic technologies for yeast replicative lifespan studies. Mech. Ageing Dev. *161*, 262–269.

156. Crane, M. M., Chen, K. L., Blue, B. W., and Kaeberlein, M. (2020). Trajectories of aging: how systems biology in yeast can illuminate mechanisms of personalized aging. Proteomics *20*, e1800420.

157. Smeal, T., Claus, J., Kennedy, B., Cole, F., and Guarente, L. (1996). Loss of transcriptional silencing causes sterility in old mother cells of S. cerevisiae. Cell *84*, 633–642.

158. Hendrickson, D. G., Soifer, I., Wranik, B. J., Kim, G., Robles, M., Gibney, P. A., and McIsaac, R. S. (2018). A new experimental platform facilitates assessment of the transcriptional and chromatin landscapes of aging yeast. Elife 7.

159. López-Otín, C., Blasco, M. A., Partridge, L., Serrano, M., and Kroemer, G. (2013). The hallmarks of aging. Cell *153*, 1194–1217.

160. Janssens, G. E., and Veenhoff, L. M. (2016). Evidence for the hallmarks of human aging in replicatively aging yeast. Microb. Cell *3*, 263–274.

161. Burtner, C. R., and Kennedy, B. K. (2010). Progeria syndromes and ageing: what is the connection? Nat. Rev. Mol. Cell Biol. *11*, 567–578.

162. McVey, M., Kaeberlein, M., Tissenbaum, H. A., and Guarente, L. (2001). The short life span of Saccharomyces cerevisiae sgs1 and srs2 mutants is a composite of normal aging processes and mitotic arrest due to defective recombination. Genetics *157*, 1531–1542.

163. Kaeberlein, M., McVey, M., and Guarente, L. (1999). The SIR2/3/4 complex and SIR2 alone promote longevity in Saccharomyces cerevisiae by two different mechanisms. Genes Dev. *13*, 2570–2580.

164. Mostoslavsky, R., Chua, K. F., Lombard, D. B., Pang, W. W., Fischer, M. R., Gellon, L., Liu, P.,

Mostoslavsky, G., Franco, S., Murphy, M. M., et al. (2006). Genomic instability and aging-like phenotype in the absence of mammalian SIRT6. Cell *124*, 315–329.

165. Kanfi, Y., Naiman, S., Amir, G., Peshti, V., Zinman, G., Nahum, L., Bar-Joseph, Z., and Cohen, H. Y. (2012). The sirtuin SIRT6 regulates lifespan in male mice. Nature *483*, 218–221.

166. Harman, D. (1956). Aging: a theory based on free radical and radiation chemistry. J Gerontol *11*, 298–300.

167. Szilard, L. (1959). On the nature of the aging process. Proc. Natl. Acad. Sci. USA *45*, 30–45.

168. Edgar, D., Shabalina, I., Camara, Y., Wredenberg, A., Calvaruso, M. A., Nijtmans, L., Nedergaard, J., Cannon, B., Larsson, N.-G., and Trifunovic, A. (2009). Random point mutations with major effects on protein-coding genes are the driving force behind premature aging in mtDNA mutator mice. Cell Metab. *10*, 131–138.

169. Hiona, A., Sanz, A., Kujoth, G. C., Pamplona, R., Seo, A. Y., Hofer, T., Someya, S., Miyakawa, T., Nakayama, C., Samhan-Arias, A. K., et al. (2010). Mitochondrial DNA mutations induce mitochondrial dysfunction, apoptosis and sarcopenia in skeletal muscle of mitochondrial DNA mutator mice. PLoS One *5*, e11468.

170. Srivastava, S. (2017). The Mitochondrial Basis of Aging and Age-Related Disorders. Genes (Basel) *8*.

171. Jazwinski, S. M. (2013). The retrograde response: when mitochondrial quality control is not enough. Biochim. Biophys. Acta *1833*, 400–409.

172. Arnould, T., Michel, S., and Renard, P. (2015). Mitochondria retrograde signaling and the UPR mt: where are we in mammals? Int. J. Mol. Sci. *16*, 18224–18251.

173. Jovaisaite, V., and Auwerx, J. (2015). The mitochondrial unfolded protein response—synchronizing genomes. Curr. Opin. Cell Biol. *33*, 74–81.

174. Liu, Z., Sekito, T., Spírek, M., Thornton, J., and Butow, R. A. (2003). Retrograde signaling is regulated by the dynamic interaction between Rtg2p and Mks1p. Mol. Cell *12*, 401–411.

175. Giannattasio, S., Liu, Z., Thornton, J., and Butow, R. A. (2005). Retrograde response to mitochondrial dysfunction is separable from TOR1/2 regulation of retrograde gene expression. J. Biol. Chem. *280*, 42528–42535.

176. Merry, B. J., and Holehan, A. M. (1985). The endocrine response to dietary restriction in the rat. Basic Life Sci. *35*, 117–141.

177. Beauchene, R. E., Bales, C. W., Bragg, C. S., Hawkins, S. T., and Mason, R. L. (1986). Effect of age of initiation of feed restriction on growth, body composition, and longevity of rats. J Gerontol *41*, 13–19.

178. Weindruch, R., and Walford, R. L. (1982). Dietary restriction in mice beginning at 1 year of age: effect on life-span and spontaneous cancer incidence. Science *215*, 1415–1418.

179. Jiang, J. C., Jaruga, E., Repnevskaya, M. V., and Jazwinski, S. M. (2000). An intervention resembling caloric restriction prolongs life span and retards aging in yeast. FASEB J. *14*, 2135–2137.

180. Medvedik, O., Lamming, D. W., Kim, K. D., and Sinclair, D. A. (2007). MSN2 and MSN4 link calorie restriction and TOR to sirtuin-mediated lifespan extension in Saccharomyces cerevisiae. PLoS Biol. *5*, e261.

181. Kaeberlein, M., Powers, R. W., Steffen, K. K., Westman, E. A., Hu, D., Dang, N., Kerr, E. O., Kirkland, K. T., Fields, S., and Kennedy, B. K. (2005). Regulation of yeast replicative life span by TOR and Sch9 in response to nutrients. Science *310*, 1193–1196.

182. Harrison, D. E., Strong, R., Sharp, Z. D., Nelson, J. F., Astle, C. M., Flurkey, K., Nadon, N. L., Wilkinson, J. E., Frenkel, K., Carter, C. S., et al. (2009). Rapamycin fed late in life extends lifespan in genetically heterogeneous mice. Nature *460*, 392–395.

183. Selman, C., Tullet, J. M. A., Wieser, D., Irvine, E., Lingard, S. J., Choudhury, A. I., Claret, M., Al-Qassab, H., Carmignac, D., Ramadani, F., et al. (2009). Ribosomal protein S6 kinase 1 signaling regulates mammalian life span. Science *326*, 140–144.

184. Bjedov, I., Toivonen, J. M., Kerr, F., Slack, C., Jacobson, J., Foley, A., and Partridge, L. (2010). Mechanisms of life span extension by rapamycin in the fruit fly Drosophila melanogaster. Cell Metab. *11*, 35–46.

185. Fontana, L., Partridge, L., and Longo, V. D. (2010). Extending healthy life span--from yeast to humans. Science *328*, 321–326.

186. Robida-Stubbs, S., Glover-Cutter, K., Lamming, D. W., Mizunuma, M., Narasimhan, S. D., Neumann-Haefelin, E., Sabatini, D. M., and Blackwell, T. K. (2012). TOR signaling and rapamycin influence longevity by regulating SKN-1/Nrf and DAF-16/FoxO. Cell Metab. *15*, 713–724.

187. Brunn, G. J., Hudson, C. C., Sekulić, A., Williams, J. M., Hosoi, H., Houghton, P. J., Lawrence, J. C., and Abraham, R. T. (1997). Phosphorylation of the translational repressor PHAS-I by the mammalian target of rapamycin. Science *277*, 99–101.

188. Cosentino, G. P., Schmelzle, T., Haghighat, A., Helliwell, S. B., Hall, M. N., and Sonenberg, N. (2000). Eap1p, a novel eukaryotic translation initiation factor 4E-associated protein in Saccharomyces cerevisiae. Mol. Cell. Biol. *20*, 4604–4613.

189. Harris, T. E., and Lawrence, J. C. (2003). TOR signaling. Sci STKE *2003*, re15.

190. Shamji, A. F., Nghiem, P., and Schreiber, S. L. (2003). Integration of growth factor and nutrient signaling: implications for cancer biology. Mol. Cell *12*, 271–280.

191. Smith, E. D., Tsuchiya, M., Fox, L. A., Dang, N., Hu, D., Kerr, E. O., Johnston, E. D., Tchao, B. N., Pak, D. N., Welton, K. L., et al. (2008). Quantitative evidence for conserved longevity pathways between divergent eukaryotic species. Genome Res. *18*, 564–570.

192. Chiocchetti, A., Zhou, J., Zhu, H., Karl, T., Haubenreisser, O., Rinnerthaler, M., Heeren, G., Oender, K., Bauer, J., Hintner, H., et al. (2007). Ribosomal proteins Rpl10 and Rps6 are potent regulators of yeast replicative life span. Exp. Gerontol. *42*, 275–286.

193. Steffen, K. K., MacKay, V. L., Kerr, E. O., Tsuchiya, M., Hu, D., Fox, L. A., Dang, N., Johnston, E. D., Oakes, J. A., Tchao, B. N., et al. (2008). Yeast life span extension by depletion of 60s ribosomal subunits is mediated by Gcn4. Cell *133*, 292–302.

194. Mittal, N., Guimaraes, J. C., Gross, T., Schmidt, A., Vina-Vilaseca, A., Nedialkova, D. D., Aeschimann, F., Leidel, S. A., Spang, A., and Zavolan, M. (2017). The Gcn4 transcription factor reduces protein synthesis capacity and extends yeast lifespan. Nat. Commun. *8*, 457.

195. Perić, M., Lovrić, A., Šarić, A., Musa, M., Bou Dib, P., Rudan, M., Nikolić, A., Sobočanec, S., Mikecin, A.-M., Dennerlein, S., et al. (2017). TORC1-mediated sensing of chaperone activity alters glucose metabolism and extends lifespan. Aging Cell *16*, 994–1005.

196. Powers, E. T., Morimoto, R. I., Dillin, A., Kelly, J. W., and Balch, W. E. (2009). Biological and chemical approaches to diseases of proteostasis deficiency. Annu. Rev. Biochem. *78*, 959–991.

197. Koga, H., Kaushik, S., and Cuervo, A. M. (2011). Protein homeostasis and aging: The importance of exquisite quality control. Ageing Res Rev *10*, 205–215.

198. Hsu, A.-L., Murphy, C. T., and Kenyon, C. (2003). Regulation of aging and age-related disease by DAF-16 and heat-shock factor. Science *300*, 1142–1145.

199. Chiang, W.-C., Ching, T.-T., Lee, H. C., Mousigian, C., and Hsu, A.-L. (2012). HSF-1 regulators DDL-1/2 link insulin-like signaling to heat-shock responses and modulation of longevity. Cell *148*, 322–334.

200. Parzych, K. R., and Klionsky, D. J. (2014). An overview of autophagy: morphology, mechanism, and regulation. Antioxid. Redox Signal. *20*, 460–473.

201. Cuervo, A. M., Bergamini, E., Brunk, U. T., Dröge, W., Ffrench, M., and Terman, A. (2005). Autophagy and aging: the importance of maintaining "clean" cells. Autophagy *1*, 131–140.

202. Cuervo, A. M. (2008). Autophagy and aging: keeping that old broom working. Trends Genet. *24*, 604–612.

203. Pyo, J.-O., Yoo, S.-M., Ahn, H.-H., Nah, J., Hong, S.-H., Kam, T.-I., Jung, S., and Jung, Y.-K. (2013). Overexpression of Atg5 in mice activates autophagy and extends lifespan. Nat. Commun. *4*, 2300.

204. Meléndez, A., Tallóczy, Z., Seaman, M., Eskelinen, E.-L., Hall, D. H., and Levine, B. (2003). Autophagy genes are essential for dauer development and life-span extension in C. elegans. Science *301*, 1387–1391.

205. Hars, E. S., Qi, H., Ryazanov, A. G., Jin, S., Cai, L., Hu, C., and Liu, L. F. (2007). Autophagy regulates ageing in C. elegans. Autophagy *3*, 93–95.

206. Kingsbury, J. M., Sen, N. D., Maeda, T., Heitman, J., and Cardenas, M. E. (2014). Endolysosomal membrane trafficking complexes drive nutrient-dependent TORC1 signaling to control cell growth in Saccharomyces cerevisiae. Genetics *196*, 1077–1089.

207. Noda, T., and Ohsumi, Y. (1998). Tor, a phosphatidylinositol kinase homologue, controls autophagy in yeast. J. Biol. Chem. *273*, 3963–3966.

208. Tylor, J. K., and Johnson, J. C. (2010). The role of autophagy in the regulation of yeast life span. Ann. N. Y. Acad. Sci. *1418*, 31–43.

209. Hu, Z., Xia, B., Postnikoff, S. D., Shen, Z.-J., Tomoiaga, A. S., Harkness, T. A., Seol, J. H., Li, W., Chen, K., and Tyler, J. K. (2018). Ssd1 and Gcn2 suppress global translation efficiency in replicatively aged yeast while their activation extends lifespan. Elife 7.

210. Li, Z., Jiao, Y., Fan, E. K., Scott, M. J., Li, Y., Li, S., Billiar, T. R., Wilson, M. A., Shi, X., and Fan, J. (2017). Aging-Impaired Filamentous Actin Polymerization Signaling Reduces Alveolar Macrophage Phagocytosis of Bacteria. J. Immunol. *199*, 3176–3186.

211. Garcia, G. G., and Miller, R. A. (2011). Age-related defects in the cytoskeleton signaling pathways of CD4 T cells. Ageing Res Rev *10*, 26–34.

212. Garcia, G. G., and Miller, R. A. (2003). Age-related defects in CD4+ T cell activation reversed by glycoprotein endopeptidase. Eur. J. Immunol. *33*, 3464–3472.

213. Garcia, G. G., and Miller, R. A. (2001). Single-cell analyses reveal two defects in peptide-specific activation of naive T cells from aged mice. J. Immunol. *166*, 3151–3157.

214. Garcia, G. G., and Miller, R. A. (2002). Age-dependent defects in TCR-triggered cytoskeletal rearrangement in CD4+ T cells. J. Immunol. *169*, 5021–5027.

215. Tamir, A., Eisenbraun, M. D., Garcia, G. G., and Miller, R. A. (2000). Age-dependent alterations in the assembly of signal transduction complexes at the site of T cell/APC interaction. J. Immunol. *165*, 1243–1251.

216. Kasper, G., Mao, L., Geissler, S., Draycheva, A., Trippens, J., Kühnisch, J., Tschirschmann, M., Kaspar, K., Perka, C., Duda, G. N., et al. (2009). Insights into mesenchymal stem cell aging: involvement of antioxidant defense and actin cytoskeleton. Stem Cells *27*, 1288–1297.

217. Higuchi-Sanabria, R., Paul, J. W., Durieux, J., Benitez, C., Frankino, P. A., Tronnes, S. U., Garcia, G., Daniele, J. R., Monshietehadi, S., and Dillin, A. (2018). Spatial regulation of the actin cytoskeleton by HSF-1 during aging. Mol. Biol. Cell *29*, 2522–2527.

218. Baird, N. A., Douglas, P. M., Simic, M. S., Grant, A. R., Moresco, J. J., Wolff, S. C., Yates, J. R., Manning, G., and Dillin, A. (2014). HSF-1-mediated cytoskeletal integrity determines thermotolerance and life span. Science *346*, 360–363.

219. Rao, K. M. (1986). Age-related decline in ligand-induced actin polymerization in human leukocytes and platelets. J Gerontol *41*, 561–566.

220. Rao, K. M., Currie, M. S., Padmanabhan, J., and Cohen, H. J. (1992). Age-related alterations in actin cytoskeleton and receptor expression in human leukocytes. J Gerontol *47*, B37-44.

221. Piazzolla, G., Tortorella, C., Serrone, M., Jirillo, E., and Antonaci, S. (1998). Modulation of cytoskeleton assembly capacity and oxidative response in aged neutrophils. Immunopharmacol Immunotoxicol *20*, 251–266.

222. Noble, J. M., Ford, G. A., and Thomas, T. H. (1999). Effect of aging on CD11b and CD69 surface expression by vesicular insertion in human polymorphonuclear leucocytes. Clin. Sci. *97*, 323–329.

223. Plackett, T. P., Boehmer, E. D., Faunce, D. E., and Kovacs, E. J. (2004). Aging and innate immune cells. J. Leukoc. Biol. *76*, 291–299.

224. Brock, M. A., and Chrest, F. (1993). Differential regulation of actin polymerization following activation of resting T lymphocytes from young and aged mice. J. Cell Physiol. *157*, 367–378.

225. Li, X., and Larsson, L. (1996). Maximum shortening velocity and myosin isoforms in single muscle fibers from young and old rats. Am. J. Physiol. *270*, C352-60.

226. Larsson, L., Li, X., and Frontera, W. R. (1997). Effects of aging on shortening velocity and myosin isoform composition in single human skeletal muscle cells. Am. J. Physiol. *272*, C638-49.

227. Degens, H., Yu, F., Li, X., and Larsson, L. (1998). Effects of age and gender on shortening velocity and myosin isoforms in single rat muscle fibres. Acta Physiol Scand *163*, 33–40.

228. Yu, F., Degens, H., Li, X., and Larsson, L. (1998). Gender- and age-related differences in the regulatory influence of thyroid hormone on the contractility and myosin composition of single rat soleus muscle fibres. Pflugers Arch. *437*, 21–30.

229. Höök, P., Sriramoju, V., and Larsson, L. (2001). Effects of aging on actin sliding speed on myosin

from single skeletal muscle cells of mice, rats, and humans. Am. J. Physiol. Cell Physiol. *280*, C782-8.

230. Ayscough, K. R., Stryker, J., Pokala, N., Sanders, M., Crews, P., and Drubin, D. G. (1997). High rates of actin filament turnover in budding yeast and roles for actin in establishment and maintenance of cell polarity revealed using the actin inhibitor latrunculin-A. J. Cell Biol. *137*, 399–416.

231. Belmont, L. D., and Drubin, D. G. (1998). The yeast V159N actin mutant reveals roles for actin dynamics in vivo. J. Cell Biol. *142*, 1289–1299.

232. Hanson, G. T., Aggeler, R., Oglesbee, D., Cannon, M., Capaldi, R. A., Tsien, R. Y., and Remington, S. J. (2004). Investigating mitochondrial redox potential with redox-sensitive green fluorescent protein indicators. J. Biol. Chem. *279*, 13044–13053.

233. Vevea, J. D., Wolken, D. M. A., Swayne, T. C., White, A. B., and Pon, L. A. (2013). Ratiometric biosensors that measure mitochondrial redox state and ATP in living yeast cells. J. Vis. Exp.

234. Chou, M. M., and Blenis, J. (1995). The 70 kDa S6 kinase: regulation of a kinase with multiple roles in mitogenic signalling. Curr. Opin. Cell Biol. *7*, 806–814.

235. Kim, E., Goraksha-Hicks, P., Li, L., Neufeld, T. P., and Guan, K.-L. (2008). Regulation of TORC1 by Rag GTPases in nutrient response. Nat. Cell Biol. *10*, 935–945.

236. Sancak, Y., Peterson, T. R., Shaul, Y. D., Lindquist, R. A., Thoreen, C. C., Bar-Peled, L., and Sabatini, D. M. (2008). The Rag GTPases bind raptor and mediate amino acid signaling to mTORC1. Science *320*, 1496–1501.

237. Panchaud, N., Péli-Gulli, M.-P., and De Virgilio, C. (2013). Amino acid deprivation inhibits TORC1 through a GTPase-activating protein complex for the Rag family GTPase Gtr1. Sci. Signal. *6*, ra42.

238. Barbet, N. C., Schneider, U., Helliwell, S. B., Stansfield, I., Tuite, M. F., and Hall, M. N. (1996). TOR controls translation initiation and early G1 progression in yeast. Mol. Biol. Cell *7*, 25–42.

239. Funchal, C., Gottfried, C., de Almeida, L. M. V., dos Santos, A. Q., Wajner, M., and Pessoa-Pureur, R. (2005). Morphological alterations and cell death provoked by the branched-chain alpha-amino acids accumulating in maple syrup urine disease in astrocytes from rat cerebral cortex. Cell Mol. Neurobiol. *25*, 851–867.

240. Juricic, P., Grönke, S., and Partridge, L. (2020). Branched-Chain Amino Acids Have Equivalent Effects to Other Essential Amino Acids on Lifespan and Aging-Related Traits in Drosophila. J. Gerontol. A, Biol. Sci. Med. Sci. *75*, 24–31.

241. Laughery, M. F., Hunter, T., Brown, A., Hoopes, J., Ostbye, T., Shumaker, T., and Wyrick, J. J. (2015). New vectors for simple and streamlined CRISPR-Cas9 genome editing in Saccharomyces cerevisiae. Yeast *32*, 711–720.

242. Higuchi-Sanabria, R., Swayne, T. C., Boldogh, I. R., and Pon, L. A. (2016). Imaging of the actin cytoskeleton and mitochondria in fixed budding yeast cells. Methods Mol. Biol. *1365*, 63–81.

243. Liao, P.-C., Wolken, D. M. A., Serrano, E., Srivastava, P., and Pon, L. A. (2020). Mitochondria-Associated Degradation Pathway (MAD) Function beyond the Outer Membrane. Cell Rep. *32*, 107902.

244. Supek, F., Bošnjak, M., Škunca, N., and Šmuc, T. (2011). REVIGO summarizes and visualizes long lists of gene ontology terms. PLoS One *6*, e21800.

245. Pernice, W. M., Vevea, J. D., and Pon, L. A. (2016). A role for Mfb1p in region-specific anchorage of high-functioning mitochondria and lifespan in Saccharomyces cerevisiae. Nat. Commun. *7*, 10595.

246. Swayne, T. C., Zhou, C., Boldogh, I. R., Charalel, J. K., McFaline-Figueroa, J. R., Thoms, S., Yang, C., Leung, G., McInnes, J., Erdmann, R., et al. (2011). Role for cER and Mmr1p in anchorage of mitochondria at sites of polarized surface growth in budding yeast. Curr. Biol. *21*, 1994–1999.

247. Adams, A. E., Botstein, D., and Drubin, D. G. (1989). A yeast actin-binding protein is encoded by SAC6, a gene found by suppression of an actin mutation. Science *243*, 231–233.

248. Miao, Y., Han, X., Zheng, L., Xie, Y., Mu, Y., Yates, J. R., and Drubin, D. G. (2016). Fimbrin phosphorylation by metaphase Cdk1 regulates actin cable dynamics in budding yeast. Nat. Commun. *7*, 11265.

249. Tkach, J. M., Yimit, A., Lee, A. Y., Riffle, M., Costanzo, M., Jaschob, D., Hendry, J. A., Ou, J., Moffat, J., Boone, C., et al. (2012). Dissecting DNA damage response pathways by analysing protein localization and abundance changes during DNA replication stress. Nat. Cell Biol. *14*, 966–976.

250. Asakura, T., Sasaki, T., Nagano, F., Satoh, A., Obaishi, H., Nishioka, H., Imamura, H., Hotta, K., Tanaka, K., Nakanishi, H., et al. (1998). Isolation and characterization of a novel actin filament-binding protein from Saccharomyces cerevisiae. Oncogene *16*, 121–130.

251. Moseley, J. B., and Goode, B. L. (2005). Differential activities and regulation of Saccharomyces cerevisiae formin proteins Bni1 and Bnr1 by Bud6. J. Biol. Chem. *280*, 28023–28033.

252. Liu, B., Larsson, L., Caballero, A., Hao, X., Oling, D., Grantham, J., and Nyström, T. (2010). The polarisome is required for segregation and retrograde transport of protein aggregates. Cell *140*, 257–267.

253. Adams, A. E., and Pringle, J. R. (1984). Relationship of actin and tubulin distribution to bud growth in wild-type and morphogenetic-mutant Saccharomyces cerevisiae. J. Cell Biol. *98*, 934–945.

254. Goley, E. D., and Welch, M. D. (2006). The ARP2/3 complex: an actin nucleator comes of age. Nat. Rev. Mol. Cell Biol. *7*, 713–726.

255. Pruyne, D., Legesse-Miller, A., Gao, L., Dong, Y., and Bretscher, A. (2004). Mechanisms of polarized growth and organelle segregation in yeast. Annu. Rev. Cell Dev. Biol. *20*, 559–591.

256. Bubnell, J., Pfister, P., Sapar, M. L., Rogers, M. E., and Feinstein, P. (2013). β2 adrenergic receptor fluorescent protein fusions traffic to the plasma membrane and retain functionality. PLoS One *8*, e74941.

257. Gauss, R., Trautwein, M., Sommer, T., and Spang, A. (2005). New modules for the repeated internal and N-terminal epitope tagging of genes in Saccharomyces cerevisiae. Yeast *22*, 1–12.

258. Gueldener, U., Heinisch, J., Koehler, G. J., Voss, D., and Hegemann, J. H. (2002). A second set of loxP marker cassettes for Cre-mediated multiple gene knockouts in budding yeast. Nucleic Acids Res. *30*, e23.

259. Longtine, M. S., McKenzie, A., Demarini, D. J., Shah, N. G., Wach, A., Brachat, A., Philippsen, P., and Pringle, J. R. (1998). Additional modules for versatile and economical PCR-based gene deletion and modification in Saccharomyces cerevisiae. Yeast *14*, 953–961.

260. Slubowski, C. J., Funk, A. D., Roesner, J. M., Paulissen, S. M., and Huang, L. S. (2015). Plasmids for C-terminal tagging in Saccharomyces cerevisiae that contain improved GFP proteins, Envy and Ivy. Yeast *32*, 379–387.

261. Rodrigues, F., van Hemert, M., Steensma, H. Y., Côrte-Real, M., and Leão, C. (2001). Red fluorescent protein (DsRed) as a reporter in Saccharomyces cerevisiae. J. Bacteriol. *183*, 3791–3794.

262. Janke, C., Magiera, M. M., Rathfelder, N., Taxis, C., Reber, S., Maekawa, H., Moreno-Borchart, A., Doenges, G., Schwob, E., Schiebel, E., et al. (2004). A versatile toolbox for PCR-based tagging of yeast genes: new fluorescent proteins, more markers and promoter substitution cassettes. Yeast *21*, 947–962.

263. Sheff, M. A., and Thorn, K. S. (2004). Optimized cassettes for fluorescent protein tagging in Saccharomyces cerevisiae. Yeast *21*, 661–670.

264. Young, C. L., Raden, D. L., Caplan, J. L., Czymmek, K. J., and Robinson, A. S. (2012). Cassette series designed for live-cell imaging of proteins and high-resolution techniques in yeast. Yeast *29*, 119–136.

265. Gietz, R. D., Schiestl, R. H., Willems, A. R., and Woods, R. A. (1995). Studies on the transformation of intact yeast cells by the LiAc/SS-DNA/PEG procedure. Yeast *11*, 355–360.

266. Madania, A., Dumoulin, P., Grava, S., Kitamoto, H., Schärer-Brodbeck, C., Soulard, A., Moreau, V., and Winsor, B. (1999). The Saccharomyces cerevisiae homologue of human Wiskott-Aldrich syndrome protein Las17p interacts with the Arp2/3 complex. Mol. Biol. Cell *10*, 3521–3538.

267. Kaksonen, M., Sun, Y., and Drubin, D. G. (2003). A pathway for association of receptors, adaptors, and actin during endocytic internalization. Cell *115*, 475–487.

268. Warren, D. T., Andrews, P. D., Gourlay, C. W., and Ayscough, K. R. (2002). Sla1p couples the yeast endocytic machinery to proteins regulating actin dynamics. J. Cell Sci. *115*, 1703–1715.

269. Miliaras, N. B., Park, J.-H., and Wendland, B. (2004). The function of the endocytic scaffold protein Pan1p depends on multiple domains. Traffic *5*, 963–978.

270. Morishita, M., and Engebrecht, J. (2005). End3p-mediated endocytosis is required for spore wall formation in Saccharomyces cerevisiae. Genetics *170*, 1561–1574.

271. Sun, Y., Carroll, S., Kaksonen, M., Toshima, J. Y., and Drubin, D. G. (2007). PtdIns(4,5)P2 turnover is required for multiple stages during clathrin- and actin-dependent endocytic internalization. J. Cell Biol. *177*, 355–367.

272. Sun, Y., Martin, A. C., and Drubin, D. G. (2006). Endocytic internalization in budding yeast requires coordinated actin nucleation and myosin motor activity. Dev. Cell *11*, 33–46.

273. Vaduva, G., Martin, N. C., and Hopper, A. K. (1997). Actin-binding verprolin is a polarity development protein required for the morphogenesis and function of the yeast actin cytoskeleton. J. Cell Biol. *139*, 1821–1833.

274. Evangelista, M., Klebl, B. M., Tong, A. H., Webb, B. A., Leeuw, T., Leberer, E., Whiteway, M., Thomas, D. Y., and Boone, C. (2000). A role for myosin-I in actin assembly through interactions with Vrp1p, Bee1p, and the Arp2/3 complex. J. Cell Biol. *148*, 353–362.

275. Jonsdottir, G. A., and Li, R. (2004). Dynamics of yeast Myosin I: evidence for a possible role in scission of endocytic vesicles. Curr. Biol. *14*, 1604–1609.

276. Doyle, T., and Botstein, D. (1996). Movement of yeast cortical actin cytoskeleton visualized in vivo. Proc. Natl. Acad. Sci. USA *93*, 3886–3891.

277. Sekiya-Kawasaki, M., Groen, A. C., Cope, M. J. T. V., Kaksonen, M., Watson, H. A., Zhang, C., Shokat, K. M., Wendland, B., McDonald, K. L., McCaffery, J. M., et al. (2003). Dynamic phosphoregulation of the cortical actin cytoskeleton and endocytic machinery revealed by real-time chemical genetic analysis. J. Cell Biol. *162*, 765–772.

278. Boldogh, I. R., Yang, H. C., Nowakowski, W. D., Karmon, S. L., Hays, L. G., Yates, J. R., and Pon, L. A. (2001). Arp2/3 complex and actin dynamics are required for actin-based mitochondrial motility in yeast. Proc. Natl. Acad. Sci. USA *98*, 3162–3167.

279. Smith, M. G., Swamy, S. R., and Pon, L. A. (2001). The life cycle of actin patches in mating yeast. J. Cell Sci. *114*, 1505–1513.

280. Karpova, T. S., Reck-Peterson, S. L., Elkind, N. B., Mooseker, M. S., Novick, P. J., and Cooper, J. A. (2000). Role of actin and Myo2p in polarized secretion and growth of Saccharomyces cerevisiae. Mol. Biol. Cell *11*, 1727–1737.

281. Waddle, J. A., Karpova, T. S., Waterston, R. H., and Cooper, J. A. (1996). Movement of cortical actin patches in yeast. J. Cell Biol. *132*, 861–870.

282. Winder, S. J., Jess, T., and Ayscough, K. R. (2003). SCP1 encodes an actin-bundling protein in yeast. Biochem. J. *375*, 287–295.

283. Okreglak, V., and Drubin, D. G. (2007). Cofilin recruitment and function during actin-mediated endocytosis dictated by actin nucleotide state. J. Cell Biol. *178*, 1251–1264.

284. Riedl, J., Crevenna, A. H., Kessenbrock, K., Yu, J. H., Neukirchen, D., Bista, M., Bradke, F., Jenne, D., Holak, T. A., Werb, Z., et al. (2008). Lifeact: a versatile marker to visualize F-actin. Nat. Methods *5*, 605–607.

285. Schneider, C. A., Rasband, W. S., and Eliceiri, K. W. (2012). NIH Image to ImageJ: 25 years of image analysis. Nat. Methods *9*, 671–675.

286. Wallace, W., Schaefer, L. H., and Swedlow, J. R. (2001). A workingperson's guide to deconvolution in light microscopy. BioTechniques *31*, 1076–8, 1080, 1082 passim.

287. Gustafsson, M. G. (2000). Surpassing the lateral resolution limit by a factor of two using structured illumination microscopy. J. Microsc. *198*, 82–87.

288. Delaney, J. R., Ahmed, U., Chou, A., Sim, S., Carr, D., Murakami, C. J., Schleit, J., Sutphin, G. L., An, E. H., Castanza, A., et al. (2013). Stress profiling of longevity mutants identifies Afg3 as a mitochondrial determinant of cytoplasmic mRNA translation and aging. Aging Cell *12*, 156–166.

289. McCormick, M. A., Delaney, J. R., Tsuchiya, M., Tsuchiyama, S., Shemorry, A., Sim, S., Chou, A. C.-Z., Ahmed, U., Carr, D., Murakami, C. J., et al. (2015). A Comprehensive Analysis of Replicative Lifespan in 4,698 Single-Gene Deletion Strains Uncovers Conserved Mechanisms of Aging. Cell Metab. *22*, 895–906.

www.ingramcontent.com/pod-product-compliance
Lightning Source LLC
LaVergne TN
LVHW042158190726
843493LV00006B/1727